BIOETHICS

BIOETHICS

A Primer for Christians

• •

SECOND EDITION

Gilbert Meilaender

William B. Eerdmans Publishing Company

Grand Rapids, Michigan / Cambridge, U.K.

First edition 1996
Second edition 2005

Wm. B. Eerdmans Publishing Co.
2140 Oak Industrial Drive N.E., Grand Rapids, Michigan 49505 /
P.O. Box 163, Cambridge CB3 9PU U.K.
www.eerdmans.com

Printed in the United States of America

12 11 10 09 08 8 7 6 5 4

Library of Congress Cataloging-in-Publication Data

ISBN 978-0-8028-2909-2

To Hannah

"Only those who are
practicing the *Tao*
will understand it."

Contents

Preface to the Second Edition

This book was first published slightly less than ten years ago, hardly a long time in world-historical terms. Yet, in the rapidly moving field of bioethics — which, as a "field," has been around for four decades at the most — ten years is a long time indeed. For several years I have felt the need to update this primer, and this second edition is the result.

Here I simply summarize for those who may wish to know the principal changes I have made from the first edition. They are, I think, four. First, the whole of the text has been updated at various points in order to make the data given more current. This is not a book primarily about data or statistics; nevertheless, there were places where it seemed important to use more recent information, and I have done so. Second, at one point — in Chapter 3 — my thinking has shifted somewhat, and I have altered the substance, or at least the tone, of the argument. In the first edition, without precisely committing myself to the view that a new individual human being comes into existence only after "twinning" has or has not occurred, I gave considerable weight to that possibility. I am now, for both empirical and "metaphysical" reasons, far less persuaded. Indeed, I think it likely that the argument that individuality is not established until approximately fourteen days of development is not going to stand the test of the embryological evidence and is likely to seem increasingly arbitrary. So I have revised the text at that point. Third, I have not only updated but also somewhat rewritten Chapter 9 (on organ donation). Although my position remains substantially unchanged, I think the tone of the argument is altered somewhat. Moreover, I have rewritten it in such a way as to take note (at least) of ongoing debates about the concept of brain death — an argument that I regard as still unfinished. Finally, I have added a new chapter in order to take up one special and highly controverted issue in research ethics: namely, research on embryos. With the possibility of the isolation of embryonic stem cells (involving destruction of embryos in the process), and with the looming possibility of cloning embryos for research purposes, this is an issue that has been of great importance in public debates. I did not think a general discussion of research ethics, without some special attention to this issue, was adequate.

The basic tone and tenor of the book remain unchanged. It is still true that this is a book written by a Christian chiefly for other Christians who want to think about some of the central issues in bioethics — though, of course, as I say in the book's introduction, others are welcome to "listen in" and consider how these issues look from the perspective of Christian vision.

INTRODUCTION

An Approach to Bioethics

The pace of medical advance in our world is so rapid that we may easily forget just how recent is the growth of bioethics as a distinct area of concern. In the minds of our children organ donation and transplantation have become facts of life; yet the first successful kidney transplant took place in 1954 and the first heart transplant as recently as 1967. Our children may assume that a pregnant woman should have the health of her fetus screened *in utero,* should know before birth whether her child is male or female, and should consider abortion for any of a number of reasons; yet, amniocentesis was first performed in 1966, and the *Roe v. Wade* decision was handed down only thirty years ago. We may assume that it is good for us to draw up advance directives about how we want to be treated if we become unconscious or otherwise incompetent, but the first living will law in this country (in the state of California) was passed in 1976. Many people simply assume that feeding tubes should be withdrawn from permanently unconscious patients; yet that step was not even seriously considered by the Quinlan family in 1976 when they went to court seeking to gain control over the medical care given their daughter Karen.

Something that might be called "medical ethics" has, of course, been present for a long time in the Western world. The Hippocratic Oath probably dates from the fourth century B.C., and physicians —

even in ages when they had none of the modern techniques for healing — reflected on the demands of their calling. After World War II, because of the deep involvement of German physicians in the Nazi regime's program of human experimentation, eugenics, and genocide, the ethics of medicine received new attention. And, indeed, the Nuremberg Code that was formulated as a response to those abuses has today the status of international law.

Only over the past three or four decades, however, has a discipline of bioethics developed, and only in these years have bioethical concerns become commonplace in our everyday lives. But as the concerns have become commonplace, they have also become the specialized possession of "bioethicists," a development that may not be wholly salutary. There was a time when philosophers and theologians, in their respective ways, thought about the moral life, and physicians reflected upon the moral meaning of their practice. Then ethics developed as a specialized branch of philosophy or theology. Now we have bioethics, one branch of what is often called "applied ethics." One of the things that happens in the course of this development is that bioethical reflection comes to focus more and more upon public policy — which in our society inevitably means a *minimal*, lowest common denominator ethic capable, it is thought, of securing public consensus. In this process, reflection upon the moral meaning of health and medicine becomes increasingly secularized, driven by the view that public consensus must exclude the larger questions about human nature and destiny that religious belief raises.

There is a place for such a minimalist bioethic, but this book aims at something different. I write as a Christian for other Christians who want to think about these issues. Anyone is, of course, welcome to "listen in" and consider what the world looks like from this angle of vision, but the discussion is not aimed at "anyone." It is aimed at those who name as Lord the God of Abraham, Isaac, and Jacob — and who believe that this Lord lived as one of us in Jesus of Nazareth. The two testaments of Christian Scripture bear witness to this God and authoritatively (even if often ambivalently) shape the vision of Christians when they turn to the contemporary concerns of bioethics. It is obvious, of course, as a matter of empirical fact, that not all Christians agree with the judgments I make in this book. But when I attempt here to write Christian ethics, I do not mean that I have taken a survey of the opinions of Christians or written a history of their views. Rather, I have tried to say what we Christians ought to say in order to be faithful to the truth that has claimed us in Jesus. A person could not attempt to speak normatively in behalf of the church unless, in Karl Barth's words, "in all humility he was willing to risk being such a Church in his own place and as well as he knew how."[1] That, I confess, I attempt here. The problems may often be new and driven by technological advance, but the search for human wisdom and faithful insight requires of us a longer memory and a more expansive vision.

1. Karl Barth, *Church Dogmatics,* I/1 (Edinburgh: T&T Clark, 1936), p. xii.

CHAPTER ONE

Christian Vision

Although a great deal of the best work in bioethics has involved the application of certain ethical principles — such as respect for autonomy, beneficence, and justice — to particular issues of concern, there is no way to apply principles in a vacuum. How we understand such principles, and how we understand the situations we encounter, will depend on background beliefs that we bring to moral reflection — beliefs about the meaning of human life, the significance of suffering and dying, and the ultimate context in which to understand our being and doing. Our views on such matters are shaped by reasoned argument and reflection less often than we like to imagine. Our background beliefs are commonly held at a kind of prearticulate level. We take them in with the air we breathe, drink them in from the surrounding culture. It is, therefore, useful sometimes to call to mind simply and straightforwardly certain basic elements in a Christian vision of the world — to remind ourselves of how contrary to the assumptions of our culture that vision may be. Hence, before we turn in the following chapters to complicated issues in bioethics, we do well to reflect briefly upon some of our background beliefs.

Individuals in Community

Bioethics talk is often talk about rights. Such talk is absolutely essential in many contexts. To ignore it is to ignore the just claims of others upon our attention and our care. But for Christians the relation of individual and community is too complex to be dealt with by such language alone, and I therefore begin with a different language.

In baptism we are handed over to God and become members of the Body of Christ. That is language about a community; yet, perhaps paradoxically, the first thing to note about baptism is that it is a deeply individualizing act. Our parents hand us over, often quite literally when sponsors carry us as infants to the font. Deeply bound as we are and always will be to our parents, we do not belong to them. In baptism God sets his hand upon us, calls us by name, and thereby establishes our uniquely individual identity and destiny.

We belong, to the whole extent of our being, only to God, whom we must learn to love even more than we love father or mother. What makes us true individuals therefore is that God calls us by name. Our individuality is not a personal achievement or power, and — most striking of all — it is established only in *community* with God. We are most ourselves not when we seek to direct and control our destiny but when we recognize and admit that our life is grounded in and sustained by God.

If the first thing to say about baptism is that it establishes our individual identity, we must immediately add that it brings us into the community of the church — with all those whom God has called by name. It is utterly impossible to exist in relation to God apart from such a bond with all others who have been baptized into Christ's Body. We are called to bear their burdens as they are called to carry ours. Sometimes we are reluctant to shoulder theirs. At least as often, perhaps, we are reluctant to have them shoulder ours, so eager are we to be masterful and independent. That others within the Body should burden us and that we should burden them is right and proper if the life of the Body is one. Nor should such mutual burdensomeness be ultimately destructive, since Jesus has been broken by these burdens once for all.

If baptism is the sacrament of initiation into Christian life, it should inform our understanding of "individualism." We should not suppose that any individual's dignity can be satisfactorily described by the language of autonomy alone — as if we were most fully human when we acted on our own, chose the course of our "life plan," or were capable and powerful enough to burden no one.

There will still remain — and should remain — a place within the political realm for the language of independent individualism. Christians should recognize that, in a world deeply disturbed by sin, great evil can be done in the name of community. Herbert Butterfield, the distinguished British historian, once suggested — only somewhat with tongue in cheek — that one could adequately explain all the wars fought in human history simply by taking the animosity present within the average church choir at any moment and giving it a history extended over time. Because sin distorts every human relationship, because, in particular, it leads the powerful to abuse and diminish the weak and voiceless in the name of high ideals or the common good, every individual's dignity must be protected. Because every person is made for God, no one is — to the whole extent of his or her being — simply a member of any human community.

Freedom and Finitude

A fuller understanding of our person requires an appreciation — and affirmation — of the created duality of our nature. That is, we are created from dust of the ground — finite beings who are limited by biological necessities and historical location. We are also free spirits, moved by the life-giving Spirit of God, created ultimately for communion with God — and therefore soaring beyond any limited understanding of our person in terms of presently "given" conditions of life.

This duality should not become a dualism, as if the person were *really* only the spirit or only the body. On the contrary, the person simply is the place where freedom and finitude are united. Body and spirit cannot be separated in our understanding of human beings; yet, because of the two-sidedness of our nature, we can look at the person from each of these angles.

Drop me from the top of a fifty-story building, and the law of gravity takes over, just as it does if we drop a stone. We are finite beings, located in space and time, subject to natural necessity. But we are also free, able sometimes to transcend the limits of nature and history. As I fall from that fifty-story building, there are truths about my experience that cannot be captured by an explanation in terms of mass and velocity. Something different happens in my fall than in the rock's fall, for this falling object is also a subject characterized by self-awareness. I can know myself as a falling object, which means that I can to some degree "distance" myself from that falling object. I cannot simply be equated with it. I am that falling object, yet I am also free from it. Likewise, I am the person constituted by the story of my life. I cannot simply be someone else with a different history. Yet, I can also, at least to some degree, step into another's story, see the world as it looks to her — and thus be free from the limits of my history. That freedom from nature and history is, finally, our freedom for God. Made for communion with God, we transcend nature and history — not in order that we may become self-creators, but in order that, acknowledging our Creator, we may recognize the true limit to human freedom.

Understanding our nature in this way, we learn something about how we should evaluate medical "progress." It cannot be acceptable simply to oppose the forward thrust of scientific medicine. That zealous desire to know, to probe the secrets of nature, to combat disease — all that is an expression of our created freedom from the limits of the "given," the freedom by which we step forth as God's representatives in the world. But a moral vision shaped by this Christian understanding of the person will also be prepared to say no to some exercises of human freedom. The never-ending project of human self-creation runs up against the limit that is God. It will always be hard to state in advance the precise boundaries that ought to limit our freedom, but we must be prepared to look for them. We must be prepared to acknowledge that there may be suffering we are free to end but ought not, that there are children who might be produced through artificial means but ought not, that there is valuable knowledge that might be gained through use of unconsenting research subjects but ought not.

In short, an ethic shaped by Christian vision will, in its general form, be what moralists term "deontological." Such an ethic does not evaluate actions only in terms of progress, only in terms of beneficial goals that might be achieved. It encourages us to exercise our freedom in search of such goals — but always within certain limits. It reminds us that others can be *wronged* even when they are not *harmed*. The only freedom worth having, a freedom that does not finally trivialize our choices, is a freedom that acknowledges its limits and does not seek to be godlike. That freedom, a truly *human* freedom, will acknowledge the duality of our nature and the limits to which it gives rise.

Person and Body

Suppose a child is born who, throughout his life, will be profoundly retarded. Or suppose an elderly woman has now become severely demented. How shall we describe such human beings? We might say, as many will today, that, although they may be living human beings, they are not persons. But we might also say — and, I think, should say — that they are severely disabled persons, the weakest among us.

It has gradually become common in our society to define personhood in terms of certain capacities. To be a person one must be conscious, self-aware, productive. The class of persons will widen or narrow depending on how many such criteria we include in our definition of personhood. But in any case the class of human beings will be wider than that of persons. Not all living human beings will qualify as persons on such a view — and, we must note, it is persons who are now regarded as bearers of rights, persons who can have interests that ought to be protected.

One might argue that such a viewpoint follows from the duality of our created nature. If the body dies, we no longer think that the living person is present. Why not reach the same conclusion if the spirit seems to have died — or never to have been present? If a human being lacks the capacities that make self-transcendence possible, why not conclude that here also the living person is not present?

The logic of this suggestion is not, however, as neat as it seems. For one thing, the duality of our nature is such that we have no access to the free spirit apart from its incarnation in the body. The living body is therefore the locus of personal presence. More important, our personal histories — precisely as histories of embodied spirits — do not require the presence of "personal" capacities throughout. Our personal histories begin in dependence — first within our mother's womb and then as newborns. Often our life also ends in the dependence of old age and the loss of capacities we once had. Personhood is not something we "have" at some point in this history. Rather, as embodied spirits or inspirited bodies, we are persons throughout the whole of that life. One whom we might baptize, one for whom we might still pray, one for whom the Spirit of Christ may still intercede "with sighs too deep for words" (Rom. 8:26) — such a one cannot be for us less than a person. Dependence is part of the story of a person's life.

Those human beings who permanently lack certain empowering cognitive capacities — as well as all human beings in stages of life where those powers are absent — are simply the weakest and most needy *members* of our community. We can care for them and about them only by acknowledging the living bodily presence that they have among us — seeking to discern in their faces the hidden spirit, the call to community that their bodily presence constitutes, and the face of Christ.

Suffering

At the heart of Christian belief lies a suffering, crucified God. Yet, in recent years some have argued that Christian emphasis upon a suffering Jesus is dangerous, that it gives rise to an ideology that encourages those who suffer oppression simply to accept that suffering. There are more things wrong with this argument than I can take up here, but it is not surprising that such arguments should arise in a culture devoted to self-realization. In such a setting, the cross must always be countercultural.

Suffering is not a good thing, not something one ought to seek for oneself or others. But it is an evil out of which the God revealed in the crucified and risen Jesus can bring good. We must therefore always be of two minds about it. We should try to care for those who suffer, but we should not imagine that suffering can be eliminated from human life or that it can have no point or purpose in our lives. Nor should we suppose that suffering must be eliminated by any means that is available to us, for a good end does not justify any and all means.

Unless we are thus of two minds, understanding suffering as an evil which can, nonetheless, have meaning and purpose, medicine is likely to go awry. It seeks *h*ealth — but not *H*ealth. The doctor is a caregiver, but not, we must remind ourselves, a savior. Ultimately, all of medicine is no more than the attempt to provide care for suffering human beings. That care, however, cannot by itself offer the Health and Wholeness we ultimately need and desire. If we respect the moral limits that ought to bind us, we will not always be able to give people what they desire. We may not be able to give the infertile couple a child, the elderly man an old age free of dependence, the young woman freedom from the child she has conceived, parents the healthy and "normal" child they had wanted, the terminally ill patient a painless death. But we can and should assure them that the story of Jesus is true — that the negative and destructive powers of the universe are not the ultimate powers whom we worship.

Part of the pain of human life is that we sometimes cannot and at other times ought not do for others what they fervently desire. Believing in the incarnation, that in Jesus God has stood with us as one of us, Christians must try to learn to stand with and beside those who suffer physically or emotionally. But that same understanding of incarnation also teaches us that to make elimination of suffering our highest priority would be to conclude mistakenly that it can have no point or purpose in our lives. We should not act as if we believe that the negative, destructive powers of the universe are finally victorious. Those who worship a crucified and risen Lord cannot give themselves over to such a vision of life.

Disease and Healing

In chapters 14–16 of 2 Chronicles we read of Asa, one of the kings of Judah. His reign, not surprisingly, was a mix of good and bad, but, in the eyes of the Chronicler, it ended badly. Rather than trusting Israel's God to bless his political aims, Asa used the temple treasure to forge an alliance with Syria in his time of need. For that he was denounced by the prophet Hanani.

A few years later Asa became severely ill; "yet," writes the Chronicler, "even in his disease he did not seek the LORD, but sought help from physicians" (2 Chron. 16:12). Shortly thereafter Asa died. The Chronicler's point — in both the political and medical examples — is clear, but it is also difficult to understand. Its clarity lies in the starkness with which we are required to ask whether the measures we take to secure ourselves — politically, medically, or in other ways — bespeak a lack of trust and confidence in God. Its difficulty, however, lies in the suggestion that God's defending and healing work is always *im*mediate, never mediated through the work of human agents.

The warning alerts us not to ask of medicine more than it can offer. Through doctors, God often treats our diseases and, sometimes perhaps, treats even our more general feeling that, although we have no identifiable disease, we are not well, are not whole. But doctors are not saviors, and the best doctors know that, even if they only think of themselves as cooperating with the powers of nature. They may heal our diseases but increase thereby our sense of invulnerability — a healing that would be disastrous for our spiritual Health. They may be unable to heal our diseases, but, accepting suffering and dependence as part of our personal history, we may be drawn closer to God. Thus there is no perfect correspondence between *h*ealth and *H*ealth.

We need not, I think, fear that seeking medical help necessarily demonstrates lack of trust or faith on our part. Rather, it indicates only that we trust God to care for us mediately — through the love and concern of others. But at the same time we should not suppose that medical caregivers can finally provide the wholeness that we need. They stand beside us, but they have not voluntarily shared our fate. They are lordly and awesome in their technical prowess, but they are not the Lord whom death could not hold.

CHAPTER TWO

Procreation versus Reproduction

Oliver O'Donovan entitled a book about artificial reproduction (or, as it is commonly termed today, "assisted reproduction") *Begotten or Made?* — a title that ought to resonate in the minds of Christians accustomed to the Nicene Creed's affirmation that Jesus Christ, the Son of the Father, was from eternity "begotten not made." And, reflecting more generally on the idea of "making babies," Leon Kass has written:

> Consider the views of life and the world reflected in the following different expressions to describe the process of generating new life. Ancient Israel, impressed with the phenomenon of transmission of life from father to son, used a word we translate as "begetting" or "siring." The Greeks, impressed with the springing forth of new life in the cyclical processes of generation and decay, called it *genesis*, from a root meaning "to come into being." . . . The premodern Christian English-speaking world, impressed with the world as given by a Creator, used the term "pro-creation." We, impressed with the machine and the gross national product (our own work of creation), employ a metaphor of the factory, "re-production."[1]

1. Leon R. Kass, M.D., *Toward a More Natural Science* (New York: The Free Press, 1985), p. 48.

Kass's observation alerts us to some of the deeper theological significance of advances in "reproductive technology." The shift from "procreation" to "reproduction" is in part a manifestation of human freedom to master and reshape our world. But especially when that mastery extends to the body, the place where we come to know a person, we should be alert to both creative and destructive possibilities in the exercise of our freedom.

Assisted reproduction includes today a number of different techniques. Artificial insemination, which in itself is not a recent development, can use either donor sperm (usually in our society from an anonymous donor) or sperm from the husband of the woman to be inseminated. Much more complex technically is *in vitro* fertilization. In this procedure both sperm and ovum are externalized in the laboratory, where fertilization takes place before the fertilized ovum (the embryo) is implanted in the woman's uterus. Sperm and ovum in this procedure usually come from husband and wife, though, of course, there is no technical reason why they must. IVF is increasingly combined with pre-implantation genetic diagnosis (PGD), in which newly fertilized embryos are tested for genetic or chromosomal defects before being implanted. More recently still, within the last ten years, a technique known as intracytoplasmic sperm injection (ICSI) has received prominent attention. A single sperm is selected and injected into the egg cytoplasm. Some clinics report up to a 90 percent success rate using this technique. First developed to deal with childlessness resulting from male infertility, it is now sometimes used to insure greater success even when male infertility is not the problem. It is disturbing to note, however, that this technique made its way into clinical practice with almost no information (from research) about its safety for the children so conceived.

Louise Brown, the first baby produced by means of IVF, was born in 1978, and by now more than a million children worldwide have been born by means of this technique. In the near future IVF will account for more than 1 percent of all births in the United States, and in some countries of Western Europe it already accounts for around 5 percent of births — all this despite the considerable expense it requires. Success rates in the United States hover around 30 percent, as does the rate of multiple pregnancies resulting from the use of IVF (since generally, in order to make success more likely, two or more embryos are implanted in the woman who hopes to become pregnant). It is also possible — though it gives rise to legal complications — for one woman to serve as surrogate for another. She may conceive a child by means of artificial insemination using sperm from the husband of the woman who will rear the child, or she may gestate and carry to term an embryo fertilized *in vitro* from gametes of the rearing parents. In short, it is technically possible for a child to have as many as five "parents": two genetic "parents" from whom sperm and ova come; the surrogate who serves as gestational mother; and two rearing parents (who need not be the same as the genetic "parents"). Such possibilities, in which human freedom intervenes to make choices possible, force us to reflect upon the meaning of the bond between parents and children. How important is the biological tie between the generations? How important is it that, except in emergencies, those who rear be those who have begotten? How important is it that a child be begotten, not made?

The Moral Meaning of the Biological Bond

It is surely natural for husband and wife to desire a child of "their own." We recognize the importance of this natural desire precisely because we find human meaning and personal significance in the biological bond that unites the generations. Were human nature simply freedom, simply will and choice, such a feeling would be inexplicable, such a desire irrational. To be sure, in speaking of a child of "one's own," we should use such language with great caution. Biological parenthood does not confer possession of a child, nor is it aimed primarily at the parents' own fulfillment. Children are "a heritage from the LORD," the psalmist writes (127:3). We are no more to hang on to them than the heavenly Father clings to his Son when the time comes for that Son to take our humanity fully into his own life. Biological parenthood confers not possession but a historical task — rearing, nurturing, and civilizing the generation that succeeds us.

With that qualification always in mind, however, we can and should recognize the human significance of the biological bond. It is significant in at least three ways. First, because we are not just free spirits but also animals, we need to know ourselves as embodied creatures who occupy a fixed place in the generations of humankind. Lines of kinship and descent locate and identify us, and, unless we learn to accept such a limit on our freedom, we remain alienated from our shared human nature. Such alienation is, at least in part, overcome as we learn to keep the commandment that calls upon us to honor our father and mother. It is, after all, a puzzling duty: to show gratitude for a bond in which we find ourselves without ever having chosen it. But that puzzle only mirrors a still greater mystery: that anything should exist at all. We did not create ourselves; we simply find ourselves here. G. K. Chesterton tells a story about his maternal grandfather, whose sons

> were criticising the General Thanksgiving in the Prayer-Book, and remarking that a good many people have very little reason to be thankful for their creation. And the old man, who was then so old that he hardly ever spoke at all, said suddenly out of his silence, "I should thank God for my creation if I knew I was a lost soul."[2]

To learn to affirm and give thanks for our place in lines of kinship and descent is to begin to learn how to give thanks for the mysterious gift of life. We learn to accept and rejoice in our limits, our creatureliness, and we learn gradually to relinquish the secret longing to be more than that. And we learn that before we can love everyone we must accept

2. G. K. Chesterton, *Autobiography* (London: Hutchinson & Co., 1937), p. 19.

the hard task of learning to love someone — that is, to love those given to us in the close tie of kinship.

There is a second sort of moral significance to be discerned in the biological tie that binds parents and children. That the sexual union of a man and a woman is naturally ordered toward the birth of children is, in itself, simple biological fact, but we may see in that fact a lesson to be learned. The act of love is not governed simply by the rational will; it is a passion that comes over us. Lovers experience ecstasy, which means that they go out of themselves, experiencing a pleasure that must be received as another's gift rather than the product of one's own reason and will. The child is God's "yes" to such mutual self-giving. That such self-spending should be fruitful is the deepest mystery not just of human procreation but of God's being. From eternity the Father "begets" the Son — that is, gives all that he is and has to the Son. Christians use just this language to affirm that God's own being is a community in love. In begetting we too give of ourselves and thereby form another who, though other, shares our nature and is equal to us in dignity. If, by contrast, we come to think of the child as a product of our reason and will, we have lost the deepest ground of human equality — and, perhaps as important, have missed the meaning of the human act of love.

A child who is thus begotten, not made, embodies the union of his father and mother. They have not simply reproduced themselves, nor are they merely a cause of which the child is an effect. Rather, the power of their mutual love has given rise to another who, though different from them and equal in dignity to them, manifests in his person the love that unites them. Their love-giving has been life-giving; it is truly *pro*creation. The act of love that overcame their separation and united them in "one flesh," that directed them out of themselves and toward each other, creates in the child a still larger community — a sign once again that such self-giving love is by God's blessing creative and fruitful. This close connection of marital love and procreation is a third aspect of the human, personal significance to be discerned in the "givenness" of the biological tie between the generations.

Third Parties

With this understanding of procreation as background, we can turn now to the moral problems raised by assisted reproduction. In our world there are countless ways to "have" a child, but the fact that the end "product" is the same does not mean that we have *done* the same thing. Artificial insemination, *in vitro* fertilization, and surrogacy may each involve parties other than husband and wife. Sperm and/or ovum may be donated, and, of course, in surrogacy not only ovum but womb may be donated. Even when the gametes are not donated, moral problems remain, and I will turn to those below. For the moment, however, we can concentrate solely on the moral significance of "third party" collaborators.

That such collaboration is often kept secret both from the child produced by it and from friends and other family members is itself an indication of unease — an indication to which we should pay heed. We might, of course, simply advise openness and honesty about the manner of this child's birth, and perhaps that will be better than placing deception at the heart of the parent-child bond, but our felt need for such deception points to the moral importance of lines of kinship and, even more, to the way in which a child should embody the union of husband and wife and spring from their embrace. Moreover, we may suspect that, once having set out on this path, we will find other — eugenic — temptations hard to resist. After all, when gametes from third party donors are used, medicine is no longer treating a disease of either husband or wife. Instead, what is being "treated" is the couple's desire to have a child. But if that is the focus of treatment, why stop with simply providing a child? Why not a child who meets certain specifications? We will consider genetic screening in more detail in a later chapter, but this is far from an idle speculation in the context of assisted reproduction. Couples who seek donated gametes to have children will generally have invested both dollars and years in the attempt. It should not surprise us if they want the best possible result. And, put more matter-of-factly, to the degree that the child has become a product, a certain quality control may begin to seem appropriate and even inevitable.

These are important concerns. More fundamental, though, is the fact that use of donated gametes — whether in artificial insemination by donor or in fertilization in the laboratory — destroys precisely those features that distinguish procreation from reproduction. Lines of kinship are blurred and confused; the child begins to resemble a product of our wills rather than the offspring of our passion; and the presence of the child no longer testifies to and embodies the union of her parents. Does all this matter? After all, one might respond by noting that even contraception brings an exercise of reason and will into the sexual act, and adoption might be said to blur lines of kinship and result in a child who cannot (biologically) embody the union of his adoptive parents.

The first of these objections recalls Chapter 1's discussion of human nature as finite and free. The Roman Catholic Church has, in fact, objected not only to assisted reproduction but also to contraception. And, although the historical grounds of its objection have been somewhat different, our discussion here should help us see how and why one might make such a connection. If assisted reproduction produces babies without sex, contraception makes possible sex without babies. In each case, one might argue, the sexual act of husband and wife is separated from the procreative good of marriage. Nevertheless, it cannot be right to rule out all human intervention in the procreative process — as if we were only finite beings who ought never seek to transcend and control what is naturally given. The use of reason and will to free ourselves from some of the constraints of nature is also part of our God-given nature. Some exercises of that freedom — even when they transcend the constraints of nature — are good and should be affirmed. Other exercises of our freedom, even when they bring desirable results, may override limits that ought not to be transcended.

We must frankly admit that it is often difficult to know when such a use of our freedom should be affirmed and when it should be condemned. Only with difficulty can we draw the needed lines here, and it is not altogether surprising that Christians have disagreed on the question of contraception. Even when they have approved contraception within marriage, however, Christians have not generally approved deliberately childless marriages. The approved contraception has been for the sake of children and directed toward fruitful marriages. If reason and will play a role here, they do so only in service of the procreative good of marriage. By contrast, the intervention of reason and will in assisted reproduction may often be quite different. The presence of third party collaborators means that we are not simply assisting a husband and wife to have a child who is the fruit of their union. Instead, lines of kinship are confused, and the child produced cannot be said to represent their union in the flesh.

But that, of course, only returns us to the other possible objection. Might we not say the same of adoption — that it confuses lines of kinship and makes possible a child who does not embody biologically the union of husband and wife? Indeed, we might, and such objections would be determinative if we had in mind a system in which children were simply taken from parents at birth and reared communally or reassigned to other adults for rearing. (That is, such an objection would be determinative if we were contemplating a program like that outlined by Socrates in the *Republic*.) These objections do not, however, rule out adoption as an emergency procedure that may be necessary when the best interests of severely abused and neglected — or simply unwanted — children require it. It is important to add, though, that these objections should help to place our understanding of adoption into proper focus. Its principal aim must not be to provide children for those who want them but are unable to conceive them. If we think in that way — to the degree that we already think in that way — some of the dangers of assisted reproduction will lie near at hand: Potential adoptive parents will want not just a child to care for but the best child, a certain kind of child. The aim of adoption, by contrast, should be to serve and care for some of the neediest among us. It may, of course, also prove fulfilling for couples who have been unable to have biological children, and there is no reason to object if their interests and the interests of potential adoptive children should coincide. But adoption must remain an emergency measure, aimed chiefly at caring for children whose biological parents have not, cannot, or will not care for them.

There are, then, good reasons for Christians to reject any process of assisted reproduction that involves sperm or ova donated by a third party. Even if the desire of an infertile couple to have children is laudable and their aim praiseworthy, even if we know of instances in which assisted reproduction seems to have brought happy results, it is the wrong method for achieving those results. What we *accomplish* may seem good; what we *do* is not. For in aiming at this desired accomplishment we begin to lose the sense of biological connection that is important to human life, we tempt ourselves to think of the child as the product of our rational will, and we destroy the intimate connection between the love-giving and life-giving aspects of the one-flesh marital union. We should not hesitate to regard reproduction that makes use of third party collaborators as wrong — even when the collaboration seems to be in a good cause.

Assisted Reproduction

What if no third parties are involved? What if the process of artificial insemination or in vitro fertilization uses sperm and ova of husband and wife? Is that also wrong? It is, I think, more difficult to render here such a clear-cut judgment, but there are reasons for concern even if no donor gametes are involved in artificial insemination or in vitro fertilization.

What is the nature of that concern? To some degree in artificial insemination, and to a considerably greater degree with in vitro fertilization, we make of our body an instrument to be used in the pursuit of our goals. We do not simply give ourselves bodily in the act of love, but we instrumentalize the body and use it in order to produce a child. In one sense, of course, this is an exercise of our freedom not unlike others in which we make use of objects and even other animals in the world to achieve our goals. To do so is to exercise the dominion given humankind by the Creator. Caution is needed, however, when the "object" used is the living human body, the place of personal presence. For in so instrumentalizing the body we are tempted to think of ourselves as only free spirit detached from the body. The real "I" becomes that free and unconstrained will that now exercises dominion even over the body it uses. What we risk here is a separation of person and body that demeans the body and makes of it a "thing."

It is not surprising, then, if we also come to think of the child who results from this process as a product — as made, not begotten. Such a move may not be logically necessary, but it lies near at hand. In begetting we form another with whom we share a nature equal in being and dignity. Since we do not transcend the child we have begotten, we do not give it worth and significance any more than we understand ourselves to have been given dignity by our progenitors. But if we make a child, we determine its meaning and use. Without supposing that every couple using assisted reproduction thinks in this way, we may still fear that a world in which we have learned matter-of-factly to accept the use of such techniques may be a world in which human "worth" increasingly becomes something to be achieved rather than the birthright of every child.

In vitro fertilization raises this issue in an obvious way by forcing us to contemplate the moral status of the embryo. Sperm and ova are externalized in the laboratory so that fertilization may occur and an embryo be formed before it is implanted in the womb of the mother-to-be. We are increasingly able to "screen" that embryo before it is implanted, to determine whether it is free of certain defects. We are able, that is, to consider whether this particular product of conception is one we desire, one whose worthiness for life we wish to affirm. Moreover, more than one embryo may be produced. All may be implanted, or only some. If some of the embryos are not implanted, we must ask ourselves what ought to be done with the remaining products of conception. They may be discarded, they may be frozen for future "use," they may (if the law permits what many researchers desire) be used as research subjects to improve our knowledge of the process of fertilization or to make progress in the struggle to cure certain diseases. It is hard for Christians to be content with any of these possibilities; yet, we should recognize that they are hard to separate from in vitro fertilization — and hard to separate conceptually once we have begun to allow ourselves to think of the embryo as made rather than begotten. Once again, we need not suppose that every couple setting foot on this path will come to think of their child as a product over which they must exercise quality control, but we deceive ourselves if we imagine that the routinized use of such techniques cannot and will not teach us to think about children in new and different ways.

Thus, even if we refrain from claiming that assisted reproduction without third party collaboration is wrong, we have good reason to fear some of the lessons it teaches us. Our world, which is so concerned that we not treat nature simply as an object over which we exercise dominion, is often strangely unconcerned when we objectify and instrumentalize the body. Indeed, we have already gone a long way in medicine toward losing the sense that the living body *is* the person, toward separating person and body. Assisted reproduction, however compelling and understandable its lure, leads us still further in that direction. We need to resist its lure and recognize its temptations.

Surrogacy

Probably the most well known surrogacy case in this country had its origins in New Jersey when the child who came to be known as "Baby M" was born to Mary Beth Whitehead in 1986. Mrs. Whitehead had contracted with William and Elizabeth Stern, who hoped to be rearing parents of a child conceived and gestated by her. She was to be inseminated artificially with Mr. Stern's sperm, and, in return, the Sterns paid her a fee (that would have totaled $20,000) and covered her medical expenses. She, in turn, contracted to bear the child and give it up after birth, agreeing also to behave in certain ways while pregnant (for example, not to smoke or to drink alcoholic beverages, not to attempt to form a mother-child bond with the child conceived, to undergo amniocentesis and abort the child if asked by Mr. Stern to do so).

The case made news — and made it into the New Jersey court system — when Mrs. Whitehead changed her mind and refused to give up "Baby M" after birth. The Sterns went to court, and the trial judge upheld the validity of their contract with Mrs. Whitehead, granted them full custody of the baby girl, terminated any parental rights Mrs. Whitehead might be thought to have, and permitted Mrs. Stern to become the adoptive mother of the child. On appeal, however, the New Jersey Supreme Court, while upholding some of the trial judge's rulings and while agreeing that it was in the best interests of Baby M to live with the Sterns, reversed some of the most crucial details of the trial judge's ruling. Specifically, it invalidated the contract between the Sterns and Mrs. Whitehead, and it restored Mrs. Whitehead's maternal rights, granting her a right of visitation.

A case such as this one demonstrates how muddled the parent-child relation becomes once choice enters in certain ways. Who, we might ask, is the "real" mother of Baby M? Is it Mrs. Stern, who proposes to rear the child? Is it Mrs. Whitehead, the genetic and gestational mother? And suppose, as could be possible, Mrs. Stern had donated the ova to be fertilized and then gestated by Mrs. Whitehead. Or suppose the ova had been donated by yet another woman and gestated by Mrs. Whitehead, with Mrs. Stern claiming a right to rear the child? Who then would be the baby's mother? We find ourselves trying to decide which of these "functions" gives a woman most claim to be the mother. The very fact that surrogacy leads us into such conundrums may itself be good enough reason to turn against it.

There are, of course, other circumstances in which such questions ought not be too difficult to answer. When a child who is unwanted or has been abused and neglected is adopted, I see little reason to doubt that the woman who rears that child is rightly described as his mother. But this, again, only indicates that adoption is not analogous to surrogacy. The child adopted is not conceived *in order* to be given up. He is already on the scene presenting in his person a need for care. Adoption is a procedure designed to answer that need already present. By contrast, perhaps the greatest moral difficulty with surrogacy is that the surrogate is being invited to conceive a human being as a means to satisfying someone else's desire to have a child. Clearly, the child then becomes an object, and, if money is paid the surrogate, a commodity. She makes of the child's person and of her body and its procreative powers instruments in service of others' purposes. That, Christians should be prepared to say, she ought not do.

Surrogacy is often defended on the ground that it is analogous to sperm donation, which is widely practiced in our society. That analogy will not, of course, carry weight for those persuaded, as I suggested above, that sperm donation is wrong because it introduces the collaboration of a third party into reproduction. But, in fact, even on its own terms the analogy overlooks something important about surrogacy. Consider Mrs. Stern's relation to Baby M. Would we describe it as maternal — or perhaps, in some respects, as more like a paternal bond? The relation of father and child has always been, by its very nature, somewhat more detached and abstract than the relation of mother and child. The father's connection to the child must be thought. An act of intellect is required to make the connection between his sperm and the child to be born. Motherhood, involving as it has not just the contribution of an ovum but also gestation, is quite a different experience. From this perspective, however, Mrs. Stern's relation to Baby M resembles the experience of paternity more than maternity. The bond is more distant and discontinuous, and to note that is to reflect from another angle on the countless ways in which — in order to pursue our projects — we may transform procreation into reproduction, which turns out to be quite a different experience. Surrogate motherhood, then, is a violation of human dignity — of the child, of the rearing mother, and of the gestational mother.

Having Children

We should not forget, of course, that people are most likely to have recourse to donor insemination, in vitro fertilization, or surrogacy because they desperately desire a child. We can understand and should sympathize with that desire. Indeed, for most people, having children is the most significant undertaking of their life. Certainly Christians, who revere "the Holy Family," should not underestimate the enormous human significance of the birth of a child. But Christians should also know that The Child has now come and lived among us and that all who follow him have been joined in one family as brothers and sisters. Without in any way undervaluing the presence of children, we should also be free of the idolatrous desire to have them at any cost — as our project rather than God's gift.

Certainly we should not fall into the language so common in our society of a right to reproduce. In many ways, in fact, despite the prevalence of such language, our society continues to regulate aspects of private sexual relationships. We establish health and age limitations for marriage. We prohibit incestuous and polygamous marriages. There is in principle no reason why we should not regulate the use of our procreative powers in ways that *wrong* others — for example, by producing a child for someone else's use — even if they do not seem, in the ordinary sense, to *harm* anyone.

The deeper issue, though, is not one of public regulation. It is a question about what sort of people we wish to be and ought to be. At least for Christians, procreation is primarily neither the exercise of a right nor a means of self-fulfillment. It is, by God's blessing, the internal fruition of the act of love, and it is a task undertaken at God's command for the sustaining of human life. Those who desire children but, it turns out, can have none are understandably saddened. Nevertheless, we must learn to pursue our projects in faithfulness to God's creative will. A couple unable to have children can — and should — find other ways in which their union may, as a union, turn outward and be fruitful. God blesses in many different ways, and the task he does not lay upon us may be replaced by other tasks less open to those who have children and equally significant for the care and preservation of the creation.

CHAPTER THREE

Abortion

Since the Supreme Court's *Roe v. Wade* decision in 1973, perhaps no issue in our nation's public life has been as bitterly debated as abortion. Yet, for Christians, this issue should be relatively straightforward, even though, of course, women trapped by an undesired pregnancy and tempted by the thought of an abortion may experience great personal anguish. For the moral questions surrounding abortion are, I think, less puzzling than some of the questions that arise in connection with assisted reproduction or some of the problems (to be discussed in a later chapter) connected to end-of-life decisions. We know what we ought to do, but we find it hard to act in accord with that knowledge. If many Christians have nonetheless become unsure about the morality of abortion, that lack of clarity represents our failure in the face of pressure from the surrounding culture. The power and importance of law is shown precisely in the fact that so many of us, even when we should know better, permit our moral judgments to be shaped by the current state of our law — supposing mistakenly that what the law permits must be morally permissible.

From the outset Christians opposed abortion. In his *History of European Morals,* W. E. H. Lecky, having noted that "the practice of abortion was one to which few persons in antiquity attached any deep feeling of condemnation," went on to note that "the language of the

Christians from the very beginning was widely different. With unwavering consistency and with the strongest emphasis, they denounced the practice, not simply as inhuman, but as definitely murder."[1] Critics will sometimes note that Christians did not always view early abortions as seriously as they did later abortions (after "animation"). Such observations miss the point, however. Christians used the best information they had available to determine when an individual human being came into existence, and Christian views gradually changed as understanding of fetal development increased. But at whatever point Christians believed an individual human being to have come into existence they regarded abortion as gravely sinful, permissible only in rare circumstances.

Because they have held that an individual human being comes into existence only at birth, only when he or she takes the "breath of life," Jews have not always regarded abortion as seriously as have Christians. And yet for many Jews — certainly for Orthodox Jews — abortion in most circumstances has been thought a great evil. Although they generally not only permit but even require abortion to protect the mother's life and health, they think otherwise about abortion as a form of backup contraception or about abortion for economic or lifestyle reasons (when a child is unwanted). And Jewish experience in the twentieth century has led many Jews to have grave reservations about abortion based on the fact that the fetus has defects that will make its life less worth living. Hence, although this is one point at which the hyphen must come out of "Judeo-Christian," Jews and Christians have differed chiefly because they have thought differently about when an individual human life begins. That is the first topic to which we now turn.

1. W. E. H. Lecky, *History of European Morals: From Augustus to Charlemagne* (1869; London: Longmans, Green, and Co., 1911), vol. 2, pp. 20, 22.

Abortion

The Beginning of Life

When does an individual human being come into existence? Christian answers to that question have been shaped by the interaction of biblical and theological concerns with changing philosophical and scientific understandings. Christians pondered the significance of biblical stories of Jacob and Esau, John the Baptist, and Jesus in their mothers' wombs. They considered the significance of original sin and the necessity of baptism. They reflected upon what it meant that Jesus had assumed the whole of human life. They were affected by theories that distinguished the formed from the unformed fetus, by preformationist theories which held that all parts of the organism preexist in its first germ and gradually unfold or develop over time (an ancient theory that, in its moral implications, is not unlike modern emphases upon genetic structure). They debated whether the soul of each new person was created by God from nothing and infused at some point into the embryo, or whether (according to traducianist theory) soul and body as a single entity were transmitted in the sexual act from parents to offspring. Medieval Catholic thought, so deeply shaped by Aquinas, tended to hold to a distinction between the formed and unformed fetus and to the creationist view that the soul was infused by God into the formed fetus. Over time, however, under pressure of both theological argument (e.g., traducianist views which made any distinction between stages of fetal development relatively unimportant) and changed scientific understanding, the creation of the soul was pushed back to the time of conception and the importance of any distinction between formed and unformed fetus was lessened. George H. Williams notes that "in its modification of creationism from Aquinas to Pius XII Catholic moral theology has indeed kept pretty close to the genetic and embryological facts" as understanding of those facts has gradually changed.[2] Biblical and theological reasons have directed at-

2. George H. Williams, "The Sacred Condominium," in *The Morality of Abortion:*

tention to the beginnings of life, and increased scientific knowledge has clarified the nature of those beginnings.

We cannot, I think, claim that the Bible itself establishes the point at which an individual life begins, although it surely directs our attention to the value of fetal life.

> For thou didst form my inward parts,
>   thou didst knit me together in my mother's womb.
> I praise thee, for thou art fearful and wonderful.
>   Wonderful are thy works!
> Thou knowest me right well;
>   my frame was not hidden from thee,
> when I was being made in secret,
>   intricately wrought in the depths of the earth.
> Thy eyes beheld my unformed substance;
>   in thy book were written, every one of them,
> the days that were formed for me,
>   when as yet there was none of them.
>
> (Ps. 139:13-16)

The poetic language of the psalmist is powerful, but if we ask it to pinpoint the beginning of life it will prove too much, since it speaks of the days that were formed for us before there were any of them. What the psalm does quite effectively, however, is depict a God who does not value achievement more than potential, who cares even for the weakest and least developed among us. More important still, at least for Christians, has been the christological teaching that in Jesus of Nazareth God has lived and redeemed the entirety of human life, from its very beginnings to the death toward which we all go. He has been with us in the darkness of the womb as he will be in the darkness of the tomb. How-

---

Legal and Historical Perspectives, ed. John T. Noonan, Jr. (Cambridge, MA: Harvard University Press, 1970), p. 169.

ever great or limited our capacities, however momentous or insignificant our achievements, we are all — in Paul Ramsey's suggestive phrase — "fellow fetuses."[3] Before God we have no claims or achievements of which to boast, and we can stand with confidence before God only because the whole of our life has been taken up into the death and resurrection of Jesus. We have, therefore, good theological reason to affirm the continuity of life from its earliest beginnings to its last breath.

Such concerns press Christians in the direction of identifying conception or fertilization as the point at which a new individual human being comes into existence, and one can appeal to our knowledge of human development to support such an identification. When sperm and ovum join to form the zygote, the individual's genotype is established. In it lies the uniqueness, the novelty, of the individual, and we can think of the rest of life as working out and developing what has been established in conception. Indeed, the concerns shared by many today about "genetic engineering," especially about possible alterations in the germ cells that are passed on to future generations, suggest how closely our sense of individual identity is tied to genotype. After fertilization it is hard to find any other equally decisive break in the process of development.

One possible alternative sometimes proposed is the point at which "twinning" either has or has not taken place. For the first fourteen days after fertilization it is possible for the developing blastocyst to "segment." That is, "twinning" can occur if the blastocyst divides into two (or more) of the same genotype. One might argue, therefore, that we cannot say with confidence that an *individual* human being exists before that point. This argument seems less persuasive to me than it once did — in part because its philosophical ground is doubtful, and in part because its basis in our knowledge of embryological development has become increasingly shaky.

3. Paul Ramsey, "Reference Points in Deciding about Abortion," in Noonan, ed., *The Morality of Abortion*, p. 67.

The claim that there can be no individual human organism before the time at which twinning either has or has not taken place seems to rest on the notion that if twinning occurs the developing blastocyst has split into two (which is the sort of thing that could happen to a clump of cells but not, except in science fiction, to a living human being). The philosophical difficulty with this claim, however, is that it seems impossible to know that this would be the right way to describe what happens if twinning occurs. The argument supposes that where once there was simply *X*, a collection of cells, there are now *Tom* and *Tim*. But, metaphysically, it is just as possible that where once there was *Tom* there are now *Tom* and *Tim*, and it seems unlikely that science can tell us why this should not be the case. The capacity to twin may simply be one of the characteristic capacities of a developing human organism at a particular (early) stage of its development.

Moreover, advancing knowledge of embryological development indicates that the beginnings of the mammalian body plan are laid down from the time of fertilization. The newly fertilized ovum has a top-bottom axis that sets up an equivalent axis in the embryo. Thus, for example, where the head and feet will sprout is established in the first hours after egg and sperm unite. Even the earliest embryo, it seems, is more than just a featureless collection of cells; it is an integrated, self-developing organism, capable (if all goes well) of the continued development that characterizes human life — and we are right to react with awe and wonder at the mystery of its individual existence.

Personhood

If this view of our earliest beginnings is correct, we should say that an individual human being — and a new human person — begins the course of his or her life's development when fertilization occurs. That person does not yet look much like us, but he looks very much as we did when we were that age. He lacks many of the capacities we take for granted as adults, but then so did we at that stage of our life — and so may we again at a future stage of our development.

Over the last several decades, however, the term "personhood" has often been used to deny protection to the developing fetus. The term points to a set of capacities — usually including consciousness and self-awareness, ability to feel pain, at least some minimal capacity for relationship with others, and perhaps some capacity for self-motivated activity. "Personhood" becomes something a living human being may or may not possess, and the class of persons becomes smaller — perhaps considerably smaller — than the class of living human beings. If we add that only persons have a right to have their life protected — or, at least, that only persons have a right of protection equal to ours — we have, of course, offered a justification for abortion. Fetal life may be of value, but, lacking personhood, its claim upon us cannot be sufficient to rule out abortion.

Popular as such "personhood" arguments have become, they have obvious difficulties which occur to almost everyone who considers them. In particular, these arguments often raise "slippery slope" worries for us. If they justify abortion, may they not justify considerably more as well? Given the description of personhood I outlined, there will be newborns, many senile elderly people, and some retarded people who do not qualify as "persons" and who therefore do not have as strong a claim upon us for protection and care as full-fledged persons do. And, indeed, one philosopher, pressing such an argument in an oft-cited article, once suggested that a fetus could have no more right to life than a newborn guppy.[4] Personhood arguments, exclusive rather than inclusive in their understanding of human community, seem in many ways to have turned against the long and arduous history in which we have slowly

4. The comment by Mary Anne Warren appears in her article, "On the Moral and Legal Status of Abortion." First published in *The Monist* 57 (January 1973), the article has been widely anthologized since then.

learned to value and protect — for Christians, to see Christ in — those who are "least" among us.

These, of course, are worries that might occur to anyone, but the problems go still deeper. Knowing that God has created us not simply as free spirits but as embodied creatures; knowing that in the child conceived in, carried by, and born to Mary God has taken the whole course of our bodily development into his own life; and knowing that even before we have the capacity for speech the Spirit intercedes for us, we can hardly find ourselves drawn toward the personhood argument. It is true, of course, that certain capacities and characteristics distinguish human beings from other species. But the personhood argument mistakenly assumes that these distinguishing characteristics constitute qualifications for membership in the human community. But to be a member of our community, with a claim for care equal to yours or mine, an individual need not possess these capacities. To "qualify" for membership he need only be begotten of human parents. Those who never had or who have now lost certain distinctive human capacities should not be described as nonpersons; rather, they are simply the weakest and least advantaged *members* of the human community. Like us, such a person is someone who has a history. Each of our personal histories begins with very limited capacities and may end in the same way. Personhood is not a thing we possess only at some moments in that history; we are persons throughout it.

Privacy

A pregnant woman's claim, "I can do with my body as I wish, can get an abortion if I wish," obviously becomes less persuasive once we reject the personhood argument. For if her body is now nourishing another human life equal in dignity to ours or hers, abortion will be far harder to justify. Nevertheless, in our society another kind of argument — grounded not in claims about the fetus's lack of personhood but in claims about the pregnant woman's right of privacy — has become pervasive. Indeed, this argument — couched in the language of fetal "viability" — was fundamental in the U.S. Supreme Court's 1973 *Roe v. Wade* decision. Although modified somewhat in more recent cases, the central determination of that decision remains in effect: At least until a fetus is "viable" and can live outside the mother's womb, the pregnant woman cannot be legally prohibited from obtaining an abortion.

I do not believe that this argument succeeds — at least as a general argument that a woman can never be obligated to carry a pregnancy to term. Indeed, it accepts and is based upon an individualism so thoroughgoing as to suppose that we have obligations to others only if we consent to them. However useful such a social contract model may be for understanding some features of civil society, it is a very poor model indeed for understanding a parent-child bond. But the privacy argument is useful in one respect. Whereas the personhood argument directs our attention only to the fetus, whose status it seeks to determine, this argument reminds us to pay attention to the bond between mother and child. Two individual human lives are involved here; yet they are so intimately united that we really have no analogue in our experience to describe their bond.

Is it unfair to the mother if we require her always to sustain this bond? After all, even after birth a child is dependent upon its mother for care, and we would not grant that she could kill the child in order to be relieved of *that* burden. Why should abortion — which is killing before birth — be any different? But it is different, of course, in one respect. After a child is born, the rest of us may assume some or all of the burdens of its care. Before birth we cannot. There are, as I said, no analogues in our experience to the relation between mother and fetus. So we must ask again: Is it unfair to the mother if we claim that she ought to sustain this bond, a bond she may, after all, not have desired or sought?

For those Christians who are pacifists the answer to this question will be governed by a more general opposition to the taking of human life. They will oppose abortion simply because they seek to welcome every human life and never to kill. For the rest of us, however, there are, I think, some circumstances — very limited circumstances — in which we should not deny her an abortion if she seeks it. We can be grateful that in our time such circumstances are medically quite rare. But if continued pregnancy constitutes a threat to the mother's life such that either she or her child must die, we cannot require her to build the human race by destroying herself. Nor must we simply wait to see what happens, as if God acted only through natural events and not through human agents, as if God did not use us to bring care and healing even in a world radically distorted by our sin. She may willingly risk her life for the sake of her child's, taking up in radical fashion the call to discipleship, the call to be Christ to *this* neighbor with whom she is so intimately bound. We may admire such a decision, we may seek to imitate her as she imitates Christ, but we cannot claim that hers is the only way to follow Christ. If she cannot take up that cross, we are entitled to give her life priority, recognizing that the child is, in truth, living off her. Clearly, of course, these are unusual circumstances, and we are now a long way from the privacy argument.

We should, I think, render a similar judgment in those cases — also very rare — of pregnancy resulting from forcible or incestuous intercourse. In the ordinary sense, of course, a woman's continued life is not threatened by such a pregnancy; yet, the case bears important analogies to that where lives conflict. For in this instance, even though the fetus is, of course, formally innocent, its continued existence within the woman may constitute for her an embodiment of the original attack upon her person. Formally innocent as the fetus itself is, it continues to represent in vivid form the attack the woman has suffered. Here again, of course, she may find the courage and strength to love and let live even the one whose presence embodies the attack of her enemy. But, again, we cannot claim that such a decision would be the only way to follow Christ.

These exceptions to the general Christian prohibition of abortion are genuinely exceptional. They are cases in which life conflicts with equal life either because the mother's life is threatened by continued pregnancy or because the presence of the fetus embodies, so to speak, a continuation of the attack the woman has suffered. As it is more generally applied, however, the privacy argument is unpersuasive. If we once grant that at every stage of its development the fetus is one of us, with a dignity equal to ours, we will not suppose that a pregnant woman's desire to be set free of that fetus can — apart from genuinely exceptional circumstances — give us good reason for abortion. Indeed, I suspect that those who extend the privacy argument beyond the exceptional cases are, in addition, presuming the truth of the personhood argument — presuming that the fetus is not, in fact, one of us, with claims upon us for care and protection. Moreover, when the privacy argument is used not just for exceptional cases but as a general principle to govern abortion decisions, it essentially confirms the view (often so appealing to men) that pregnancies and children are the private responsibility of women. By contrast, as I noted in the previous chapter, in the Christian vision the union of husband and wife is embodied by the child, for whom both must therefore accept responsibility.

Welcoming Children

The Christian rejection of abortion in most instances is not, however, finally grounded chiefly in a critique of the personhood and privacy arguments. Rather, we seek daily to learn how to see the whole of life in the light of God's creative and redemptive activity. The life of the child in the womb is God's creation, and that child is part of the world Christ came to redeem. The worth and dignity of the child's life are not therefore dependent on our evaluation — on whether at any given moment we "want" that child. Indeed, both before and after birth parents' feelings may vacillate — as they sometimes want and at other times do not want their children. That the rabbis understood this well is nicely captured in a few sentences by David Feldman:

> The Bible prescribes that an offering be brought to the sanctuary by the woman following childbirth (Lev. 12:6). Its purpose, explains the Talmud, is to atone for a vow never meant to be kept: When birth pangs were severe, she presumably vowed "never again"; a while later she would forget that oath; satisfaction had dispelled anxiety.[5]

Our continuing task, therefore, is to struggle to bring our judgments and feelings into accord with God's action — to let our estimate of the child be shaped and formed by God's.

Seriously to attempt this is to learn our limits. We do not, ultimately, fashion the conditions of our life; rather, we live under God's mysterious but providential governance. The unexpected — and even the unwanted — events of life are occasions and opportunities for hearing the call of God and responding faithfully. Sometimes, perhaps often, this will mean that we take up tasks and burdens we had not anticipated or desired, and they in turn may bring a certain measure of suffering. Within the community of the church, of course, we ought to seek to bear each other's burdens, and too often we fail to do so. But even when we think we suffer alone, we do not, since God has taken that suffering into his own life.

To counsel the acceptance of the unwanted — acceptance even of the suffering it brings — is not to encourage mothers or fathers to be "victims." Rather, it is to call for the strength that virtuous action requires. One need not be a Christian to agree with Socrates that it is better to suffer evil than to do it, but certainly Christians should understand such a claim. If we seek to save ourselves by doing away with the

5. David M. Feldman, *Marital Relations, Birth Control, and Abortion in Jewish Law* (New York: Schocken Books, 1974), p. 294.

child who is unwanted, we hand ourselves over to the destructive powers of the world in an attempt to avoid them, and we act as if those powers are ultimately worthy of our worship, as if they could save. We take our stand, it is sobering to realize, beside King Herod after he heard the news the Magi brought. That is not, I think, where, finally, we want to be.

CHAPTER FOUR

Genetic Advance

In 1990 an international consortium, with a sizable funding commitment from the federal government of the United States, began an initiative known as the Human Genome Project. Its goal was to provide a complete map of the human genome. Originally scheduled to extend over a fifteen-year period and to cost $3 billion, the project was completed early and somewhat under the original cost estimates. The first draft of the human genome sequence was announced in June 2000, and the announcement of the finished sequence came on April 14, 2003 (just eleven days before the fiftieth anniversary of the publication of Watson and Crick's description of the double helix structure of DNA). No doubt the scientists engaged in this work were driven partly by the simple desire to know — the urge behind all pure science. Clearly, though, the government foots a bill of this size largely for other reasons, because the initiative holds out enormous potential for better treatment of hereditary diseases.

The pace of advance over the last decade of genetic research has been astonishing. Every year brings news articles detailing the discovery of genes now known to cause one or another hereditary disease. Some 1,400 disease genes have been identified, and the list will grow. (Other inherited diseases may be caused by the combined effect of several genes or by some combination of genetic and environmental causes.) One of the future goals of researchers is to develop a technology able to sequence an individual's entire genome for a cost of less than $1,000. We have to be careful, though, in describing the kind of progress this has made possible. In a sense, we now find ourselves returning to an earlier condition of medicine. Before the discovery of antibiotics physicians could often diagnose much more than they could cure. The era of "wonder drugs" taught us to expect from doctors not only diagnosis but also treatment and cure. Now, however, at least for the time being as the era of genomic medicine dawns, we again find ourselves with knowledge of diseases for which we can offer little or no treatment. Indeed, in many instances today, the "treatment" of choice is prenatal diagnosis followed by abortion if a genetic defect is detected. That "treatment" I will take up in the next chapter. In this chapter, however, we will reflect upon some of the more general issues raised by our rapidly increasing genetic knowledge.

Genetic Therapy

Put a little too simply, we could say that we get a set of each of our genes from both our mother and our father. Usually both sets of a gene are the same, but sometimes one is a mutant gene carrying a defect. If the mutation is not lethal, it is then passed on to future generations when its carrier has children. Often this makes little difference, because the cell's function can still be controlled by the other, normal gene. But some mutant genes are known as "dominant" because they dominate their counterpart. Thus, for example, the gene that causes Huntington's disease is dominant. It will express itself even if its paired copy is normal. To carry such a dominant gene in one's cells means that one will inevitably have the disease (even though, as with Huntington's disease, for example, the disease may not express itself early in life). Other mutant genes are inherited "recessively." That is, they do not dominate their paired counterpart. Recessively inherited diseases — such as cystic fibrosis or Tay-Sachs — do not express themselves unless both copies of that gene are mutant. Thus, one could be a carrier of the gene that causes cystic fibrosis without ever having the disease. Should two such carriers have children, however, some of their children, inheriting the defective gene from both mother and father, could have the disease. Still other recessively inherited diseases, such as hemophilia, are carried by genes on the chromosome that determines sex. Male children often inherit such disorders (since their XY-chromosome cannot mask a defective X-chromosome inherited from their mother), and female children often become carriers of the disorder. Some other well known inherited disorders, such as Down's syndrome, result not from mutant genes but from chromosomal abnormalities.

Of course, not all inherited genetic disorders are as serious as these examples, but, quite obviously, therapies that offered a cure for such dreaded diseases would be a tremendous medical advance. Two types of possible genetic therapy are often distinguished — somatic cell and germ cell therapy. Germ cells are those cells that will become sperm or ova; hence, alterations in germ cells will be passed on to future generations of descendants as inheritable genetic modifications. Somatic cells (from *soma,* the Greek word for "body") are simply all bodily cells that are not involved in reproduction. Changes in somatic cells are not passed on to one's children; they die when the individual carrying them dies. Some beginning steps in somatic cell therapies have been taken in recent years. Since most such research in this country is federally funded, proposals for human gene therapy must be approved by the Recombinant Advisory Committee of the National Institutes of Health. The first such experiment was carried out in 1990, when researchers studying a rare immune disorder (ADA deficiency) used gene replacement in bone marrow to attempt to treat the disorder in a young girl. Researchers have also used gene replacement techniques to attempt to treat cystic fibrosis, hemophilia, and certain skin cancers.

Research into somatic cell therapy received a serious setback in

1999, when a young man named Jesse Gelsinger, who was participating as a research subject in an experiment at the University of Pennsylvania aimed at treating an inherited disease of the liver, suddenly died from a blood-clotting disorder caused, evidently, by the virus used as a vector to insert a corrective gene. Even apart from such tragic events, progress in somatic cell gene transfer research has been slower than had been predicted a decade or more ago. It was not until April of 2000 that researchers (in France) announced what seems to have been the first true example of success in such gene transfer research. The research subjects were infants with severe immune disorders, for whom the only available (but not very effective) treatment would have been bone marrow transplantation. Even this success story had a less than happy ending, however, for by September 2002 it had become clear that several of the young research subjects had developed a form of leukemia — apparently as a side effect of the treatment.

To accomplish germ cell rather than simply somatic cell modification, genes would have to be inserted into the germ line of an organism. This could be done by genetifically modifying the gametes themselves before fertilization, by modifying the fertilized ovum, or by gene transfer into the cells of a very early embryo in such a way as to modify its gametes. Germ cell modification has often been thought to be the stuff of science fiction, far less likely to become a genuine possibility in the immediate future. That judgment may, however, have to be revised; for late in 1994 researchers reported that they had successfully altered germ cells in mice so that the alterations were passed on to the next generation. Although a technique that succeeds with mice might fail in humans, and although there would surely be many complications and difficulties in attempting that jump, research into methods of producing inheritable genetic modifications in animals proceeds on various fronts, and it raises stunning and disturbing possibilities.

How shall we evaluate germ cell therapy? There is something awesome about an intervention that no longer deals only with the *soma,* the bodily form, but that, instead, cuts to the core of the identity not just of one person but also of his descendants. If anything amounts to "playing God" illicitly, germ cell modification might seem to. Nevertheless, germ cell therapy, were it possible, offers an obvious benefit. It treats a disease not just in one sufferer but in all of his descendants. Why should that not be preferred to continued tinkering with the *soma* of sufferers generation after generation? Would it not, in fact, be a marvelous exercise of the freedom to reshape themselves with which human beings are endowed, a freedom by which we transcend the limits of what once seemed "given"? (And, in fact, would there be any convincing reason to stop with modifications that were clearly therapeutic when the same techniques might someday enable us to "enhance" capacities that were already perfectly normal?) To draw back in fear here might seem to be the sin that was once called "sloth" — an unwillingness to seize new possibilities and a readiness to fall back into the safe and familiar.

Perhaps. But we are not only free, self-determining beings; we are also finite creatures of God. Hence, we are not self-creators, and some limits of our finitude ought to be respected. With germ cell modification we may have come upon such a boundary. To step across it might be, not simply to exercise our freedom in an awesome way, but to step across in a more fundamental sense, to transgress. Why might one think that? Precisely the supposed great benefit of germ cell therapy — that it would overcome disease not just in one person but in future generations — is its greatest danger. That danger is what C. S. Lewis memorably characterized as "the abolition of man." Himself a writer of science fiction, he contemplated exactly the sort of advance that germ cell therapy may one day offer us. What seems like an increase in our capacity to master and control nature suggested, to his mind, something quite different. "Man's power over nature turns out to be a power exercised by some men over other men with Nature as its instrument. . . . Each new power won *by* man is a power *over* man as well."[1]

1. C. S. Lewis, *The Abolition of Man* (New York: Macmillan, 1947), pp. 69, 71.

When we take up the project of shaping future generations in so fundamental a way, we cannot really know what good or ill we may accomplish — we cannot, that is, really know what project we undertake. We may even wonder whether such a project, whatever exactly it may turn out to be, is really part of medicine. For the "patient" is no longer any particular suffering human being, but humankind. What estimate of ourselves — of our virtue and wisdom — would we need even to want to become so fundamentally the shapers of humanity? That the project is alluring we cannot doubt. And such deeds, if they are done, will be the work of people like us — people who are moved by benevolent concern to relieve future suffering (as well, of course, as by the desire for knowledge, power, and fame). Yet, the most truly human and humane exercise of our freedom may be the courage that says no when asked to make humankind itself our patient.

Somatic cell therapy, by contrast, does focus on the needs of particular patients, and because it is less far-reaching in its aims it is also less controversial in principle. If, by replacing the mutant gene that causes a person to suffer from cystic fibrosis, we could successfully treat an afflicted patient, that would be an outcome well worth seeking and supporting. The moral questions raised by somatic cell therapy are slightly different, and they call not for the no that should be spoken to germ cell modification but for caution and a willingness to distinguish acceptable from unacceptable aims of therapy. We cannot predict with any certainty the future course of scientific advance, but a day may come when — even granting the complex interactions of nature and nurture — it will be possible to alter a gene or genes that strongly determine intellectual capacities. Is such an intervention different from somatic cell therapy aimed at curing Tay-Sachs disease or diabetes?

There is, I think, an important difference between treating what everyone acknowledges to be a disease and seeking to enhance intellectual capacities. This line may sometimes be difficult to draw in practice, but that means only that it is difficult to draw, not that it does not exist or is unimportant. In order to draw it at all, however, we need to resist the inclination — rather common among religious people who want to think of Health and Wholeness in wide-ranging ways — toward expansive definitions of health. As long ago as 1946 the World Health Organization offered the classic example of such an expansive definition when it described health as "a state of complete physical, mental and social well-being, and not merely the absence of disease or infirmity." That definition has been subjected to a good deal of scrutiny over the years in the bioethics literature. Although, as I noted, religious people are often inclined toward such definitions, Christians ought, in fact, to be wary of them. They threaten to turn the search for health into the all-consuming passion of life and to make the doctor the savior from whom wholeness ultimately comes. Moreover, if we accept such a definition, there will be no limit, apart from the current frontiers of medical advance, to our efforts to reshape and reform those who fall short of complete well-being. We may never reach a point at which we are prepared to accept others, and especially our children, as they are — to oppose diseases but accept persons whatever their capacities.

Parents do, of course, have a responsibility within limits to shape the lives of their children. But much of the time we may already be too eager to use the methods available to us — chiefly, for now, the control of nurture rather than nature — to carry out this task. Those who doubt this need only observe parents watching their children participate in school sports or applying to colleges. We are very reluctant to let the mystery of personhood — equal in dignity to our own — unfold in the lives of our children.

What we need really is not just a carefully circumscribed definition of health and disease. We need the virtue of humility before the mystery of human personhood and the succession of generations. We need the realization that the children who come after us are not simply a product for us to mold. This is not, I repeat, to deny that we have a responsibility to inculcate our children into all that is good and true. And as Christians we have, in particular, the duty to initiate our children into the heritage of the church. But we do that first of all when in baptism we hand the child over to God, recognizing that the faith is not simply our possession to be transmitted as we wish. As John Henry Newman once wrote: "Basil and Julian were fellow-students at the schools of Athens: and one became the Saint and Doctor of the Church, the other her scoffing and relentless foe."[2] Faith remains a gift, and its character as gift should be for us a constant reminder of the limits to the nurture we provide. Were such humility to shape our vision generally, we would, I think, have little to fear from somatic cell therapy. We could pursue its possibilities while also being more accepting of our recalcitrant genes. We could pursue it without dreams of mastery.

Screening

When we consider the possibilities for genetic therapy, we are for the most part discussing what we might, using a grammatical metaphor, characterize as the "future subjunctives" of genetic advance. Although amazing progress has been made in researching the possibilities of such therapy, it remains largely at the stage of experiment. The "present indicatives" of genetic advance, at a moment in time when we can diagnose far more than we can treat, are the several kinds of genetic screening that are presently available. The most important — and morally the most troubling — form of screening is prenatal diagnosis of the fetus still in utero. That topic will be taken up in the next chapter. Some forms of newborn screening and carrier screening are also available, however, and the latter especially deserves some attention.

Newborn screening is aimed at detecting diseases that are acquired through recessive inheritance and which, if detected soon enough, can be successfully treated. The blood sample taken by prick-

2. John Henry Cardinal Newman, *The Idea of a University,* ed. Martin J. Svaglic (Notre Dame: University of Notre Dame Press, 1982), p. 161.

ing the heel of newborn infants is the most common example of such screening. That sample is tested for PKU, a recessively inherited enzyme disorder which prevents the breakdown of phenylalanine, an amino acid, and results in severe retardation. If, however, a PKU infant is placed at once on a diet very low in phenylalanine, these consequences can be prevented. Even though the diet is a very restricted one, its benefits are so obvious that it is hard to regard PKU screening as anything other than a success.

Carrier screening is not aimed at the detection of disease itself. Rather, it tells people whether they are the carriers of mutant genes that can be inherited recessively. This is information one might well want before marrying and conceiving children. For a carrier of a recessive disease, though not himself affected by it, may mate with someone who is also a carrier. Their children will be at risk, not just of being carriers, but of having the disease.

There is striking evidence that, under special circumstances, a program of carrier screening can markedly reduce the number of children born with recessively inherited genetic disorders.[3] A community of Orthodox Jews in New York City, especially at risk for Tay-Sachs disease, developed a program called "Dor Yeshorim" ("the generation of the righteous"). It uses carrier screening to "discourage marriage or even dating" between young people who, because both are carriers of a recessively inherited disorder, would be at risk of having a child with the disease. (The program has, in fact, been extended to screen not only for Tay-Sachs but also for other conditions such as cystic fibrosis and Gaucher's disease.) Teenagers in high school are tested to determine their carrier status. That information, encoded by means of identification numbers, is recorded at a central office. If a boy and girl want to date, they are encouraged to check with the office to determine

3. For the information that follows, see Gina Kolata, "Nightmare or the Dream of a New Era in Genetics?" *The New York Times*, 7 December 1993, p. A1. See also Christian Rosen, "Eugenics — Sacred and Profane," *The New Atlantis*, no. 2 (Summer 2003): 79-89.

whether they would be at risk of having children with inherited genetic diseases. The program has greatly reduced the incidence of Tay-Sachs within this Orthodox Jewish community, not by aborting defective fetuses, but by counseling couples at risk for having afflicted children.

We may wonder whether the extension of this screening program beyond Tay-Sachs to cystic fibrosis (which, unlike Tay-Sachs, is treatable) and to Gaucher's disease (the first symptoms of which do not usually even appear before age forty-five) is wise, but we should also take its possibilities seriously. Such a program is probably not possible, and may not even be desirable, on a larger scale. Its success is due partly to the fact that the Orthodox Jewish community is so tightly knit and clear about its shared beliefs and commitments. On a larger scale the dangers of social control might be too great to risk. But Christians should have little sympathy for those who resist such control before conception while willingly screening and aborting defective fetuses after conception. Moreover, Christian belief that the birth of a child is a sign of God's continued blessing upon his creation does not mean that we should succumb to the current tendency to characterize procreation as a right. It is not principally that, nor is it an exercise in self-expression or self-fulfillment. Procreation, as the internal fruition of the act of love, is by God's blessing a task undertaken to sustain human life. That task, like all callings, should be undertaken responsibly. Doing so may sometimes lead not to fulfillment of our desires but to sacrifice of some of them. That truth we might learn once again from "Dor Yeshorim."

CHAPTER FIVE

Prenatal Screening

Of all the virtues, *love* is perhaps the one that has been most emphasized by Christians. Theologians — as well as others — have written at length, exploring its depth and complexity. If, however, we had to express in a sentence the meaning of human love as a reflection of and response to God's own love, it would be hard to do better than the formula of Josef Pieper: Love is a way of saying to another, "It's good that you exist; it's good that you are in this world!"[1] Precisely because we know ourselves to have been loved this unqualifiedly by God, and because we know we should learn to love others as we have been loved, Christians ought to set themselves against prenatal screening, at least as it is currently practiced in this country in an increasingly routinized way. For it stands in conflict with the virtue that would say to another: "It's good that you exist."

Prenatal screening of the fetus *in utero* — usually by amniocentesis, with abortion in view as a "treatment" possibility if the fetus has inherited genetic defects — has increasingly become a routine part of medical practice. As such a routinized practice it clearly stands in conflict with the virtue of love. At least for the present, while we can diag-

1. Josef Pieper, *About Love,* trans. Richard and Clara Winston (Chicago: Franciscan Herald Press, 1974), p. 19.

nose far more fetal defects than we can cure, this is the principal use of advances in genetic knowledge. And it is not, of course, treatment of a disorder; it is simply elimination of the one who is afflicted with a disorder. To that one we do *not* say: "It's good that you exist."

In order to think more deeply about this issue, I want to set before us four cases, to which I will return along the way:

(1) A pregnant woman undergoes amniocentesis to test for genetic abnormalities in the child she is carrying and is told that she will give birth to a normal male child. At term, however, she gives birth to a girl who is afflicted with Down's syndrome. The woman, contending that she would have aborted the child had she been correctly informed about the test results, sues the hospital that administered the tests — but sues, not on her own behalf, but on behalf of the child. That is, she contends that the hospital should pay damages for the child's "wrongful life." (A case something like this was heard in the New Jersey court system in the mid-1980s.)

(2) We might imagine a slightly different kind of "wrongful life" case. A thirty-seven-year-old woman gives birth to a son born with Down's syndrome and some other physical abnormalities. This woman, having been offered the possibility of amniocentesis while pregnant, had declined it. Some years later her son sues not the hospital or physician but his mother for his "wrongful life."

(3) A pregnant woman has placenta previa — a serious condition in which the placenta blocks part of the cervix, leading to a risk of hemorrhage and oxygen deprivation for the fetus. The woman's physician has advised her not to use amphetamines or have sexual intercourse with her husband for the remainder of her pregnancy. If she begins to bleed, she is to go at once to the hospital. The woman's child is born with severe brain damage and dies after six weeks. It turns out, however, that the woman had delayed going to the hospital for a number of hours after she had begun

bleeding, had been using amphetamines (taking them as recently as the day she went into labor), and had sexual relations with her husband (also on the day she went into labor). (A case similar to this one occurred in California, and the woman was charged with a misdemeanor under the state's child abuse law. The case was, however, eventually dismissed.)

(4) A pregnant woman regularly uses crack cocaine even after the fetus she is carrying is viable and could live outside the womb. Although she knows that this could harm or even kill the child, she continues, and, in the event, her unborn child dies. She, in turn, is then tried and convicted for homicide under the state's child abuse laws. (A case like this occurred in South Carolina only a few years ago.)

Cases like these are important in several respects. Here, however, I will use them to see what they may teach us about the way in which genetic screening and readily available abortion transform the meaning of parenthood. We can begin with the last two cases. They could, of course, have many variations. The pregnant woman could be a heavy smoker or drinker. She could need to go on a special diet in order to protect her fetus from certain dangers. The fetus might need surgery while still in the uterus — a kind of science fiction scenario that is actually possible in certain instances. Needing a caesarean section to deliver her child safely, the woman might not want one. Such possibilities force us to ask whether a pregnant woman could be required to alter her behavior before the birth of her child for the sake of that child's health, and whether she could be penalized after birth if an unwillingness to change her behavior had brought harm to the child.

We might initially be quite reluctant to recommend what could amount to government supervision and monitoring of many pregnancies. Yet the harm to children that we are discussing here is real, and we do, of course, punish both child abuse and neglect when they occur after birth. Although in this country a woman is at present legally free to abort her child and thereby end a pregnancy, one might argue that she is not free to carry that child to term while behaving in ways that may seriously harm it. Without at present any legal obligation to carry the pregnancy to term, she might still have legal obligations to avoid conduct that will harm her child's future. Some, however, have wanted to argue against any such legal obligation. Thus, for example, Dawn Johnsen, a legal scholar, wrote that laws requiring the pregnant woman to alter her behavior "treat the fetus as if it were an independent entity, physically separate from the woman, and ignore the fact that the woman cannot walk away from the fetus."[2]

But, in fact, she *can* walk away from it. Legal abortion makes it possible for her to do precisely that. We must notice the implications of that possibility, since freedom is not always as liberating as we anticipate. The possibility of abortion has a kind of peculiar effect upon the bond between the woman and the fetus she carries. The "natural" connection between them ceases to be as tight once "choice" enters the picture. Now she can walk away.

This looks at first as if it gives the woman greater freedom — the freedom to see herself and the fetus as separate entities, the freedom to affirm the connection between those entities if she so chooses or to reject the connection. But it now turns out, as we contemplate my third and fourth cases above, that this may not be liberating. Because the pregnant woman has not walked away, she may have taken on quite stringent obligations. Choice, freedom, brings with it some heavy burdens.

With such a possibility in mind, we can turn to the issue of prenatal screening as it is raised by the first two cases above. The sociologist Barbara Katz Rothman has offered a very insightful analysis of the complexities — psychological and moral — of prenatal diagnosis.[3]

2. Dawn Johnsen, "A New Threat to Pregnant Women's Autonomy," *Hastings Center Report* 17 (August/September 1987): 35-36.

3. Barbara Katz Rothman, *The Tentative Pregnancy: Prenatal Diagnosis and the Future of Motherhood,* 2nd ed. (New York and London: W. W. Norton, 1993). Rothman herself, it is important to emphasize, does not oppose a legal right to abortion.

Without by any means doing justice to its intricacies, I will summarize a few aspects of her argument, because it forces us to think clearly and critically about the implications of prenatal diagnosis with abortion as the possible "treatment" in view.

In several respects the technology now available encourages a mother to distance herself from the child she carries. We are able to learn a great deal about a fetus (including whether it is male or female) through genetic screening, and we can, in a sense, visualize it by means of a sonogram. Such possibilities reverse the natural connection of mother and child. Before the development of such technologies, mother and child began with an inseparable attachment and moved toward birth, the beginning point in their distancing and the separation of their individual identities. But prenatal screening means that the relation of mother and child begins with separation and distancing and moves toward birth, which is now understood as the moment of attachment or "bonding." Screening invites us, Rothman suggests, to think of pregnancy as moving from separation to attachment as we keep open the possibility that the woman may "walk away" from this pregnancy and child.

For this reason the pregnancy becomes "tentative." Unless and until the screening yields reassuring results, the pregnant woman will be reluctant to acknowledge the presence of the fetus she carries. She must maintain a certain distance, because she does not yet know whether she will sustain the bond with this fetus or whether she will walk away. Thus, prenatal screening with abortion as a possible "treatment" in view if test results are unsatisfactory has a subtle effect on the meaning of motherhood (and, eventually, also fatherhood). For it makes the commitment of mother to child tentative, conditional. We should not underestimate the symbolic force of such a qualified and conditional commitment, especially when we remember that no technology can assure parents a perfect baby. No technology can guarantee that unpredictable problems and disabilities will not arise after birth. Thus, Rothman writes: "The possibility of spending the rest of one's life caring for a sick or disabled child can *never* be eliminated by prenatal testing. I worry about women who say they only dare have children because prenatal diagnosis is available. Motherhood is, among other things, one more chance for a speeding truck to ruin your life."[4] Prenatal screening is, for this reason, poor preparation for becoming a mother or father.

Moreover, if we let our imaginations roam a little, we may be struck by the metaphysical implications of this technology. The "wrongful life" cases with which I began this chapter are a good prism through which to contemplate these deeper implications. Courts have, on the whole, been quite reluctant to permit such suits — reluctant, that is, to grant that existence itself could be an evil on the basis of which one had standing to sue — although courts in a few of our states have permitted such suits to proceed. To my knowledge, no case quite like the second above — in which a child sues his mother for "wrongful life" — has yet occurred. If we continue alone the road we are now traveling, however, there is no reason to suppose it will not.

Making only a conditional commitment to one's unborn child, holding open the possibility of walking away, seems to give a certain freedom. But here again we note that what looks like greater freedom gives rise to greater responsibility. If we create a product for certain purposes, we can be held responsible for the quality of that product. Traditionally, of course, parents conceiving a child did not think of themselves as producing a product. If the child turned out to be less than perfect, the parents could turn to God — whether in submission to his providence or in Job-like anger. In either case, they were only co-creators with God, not simply creators with ultimate responsibility for the fate of their child. As technology makes possible a more complete responsibility for the child's well-being, so it also lays upon all who use it a heavier burden of responsibility. Complete freedom, godlike freedom, gives rise to utter responsibility. "Wrongful life" suits simply rec-

4. Rothman, *The Tentative Pregnancy,* pp. 252-53.

ognize the fact that we have begun to think of ourselves not simply as cooperators with a power greater than our own but as ultimate life-givers. And then we cannot avoid the impetus toward "quality control."

There is no way out of this trap by going forward. Amniocentesis can on occasion be put to good use. (It was, in fact, first developed to detect Rh blood group incompatibilities between a pregnant woman and the child she was carrying.) But we deceive ourselves if we suppose that, as a routine feature of medical practice, it can simply assist a couple to prepare themselves for their child's birth. It does exactly the opposite. It sets our foot on a path that is difficult to exit. We may tell ourselves that we only want to know the health of the fetus, that abortion is not a possible end in view, but, for the most part, I think, we thereby deceive ourselves. The technology carries its own momentum, which, if not irresistible, is nevertheless very powerful. It prepares us not for the kind of commitment that parenthood requires, an unconditional commitment, but for a kind of responsibility that finite beings ought to reject. The time of pregnancy will be better spent learning to love the child we have been given before we begin to evaluate and assess that child's capacities. Christians could do the world a considerable favor and could bear substantial witness to the meaning of God's own love for the world if they would simply say no to routinized prenatal screening — thereby saying to their children and, by implication, to all others: "It's good that you exist."

CHAPTER SIX

Suicide and Euthanasia

Here is a story of the sort one occasionally reads in the newspaper.[1] A man named George Delury helped his wife of twenty-two years, Myrna Lebov, to commit suicide. Ms. Lebov was suffering from multiple sclerosis and near the end of her life was able only to wash her face and hands, feed herself if the food had already been cut, apply her makeup, and brush her teeth. In addition, her memory had begun to be affected. When she asked her husband if he would help her end her life, he reports that "I said, of course I'd help. I also said in effect that I was astonished she'd fought so hard and so long to keep going. I would have quit a lot sooner."

He learned that one of the medications his wife was taking could prove fatal if taken in a sufficiently large dose, so over several months she cut back on her dosage, saving enough of the medicine to kill herself. Then on Independence Day they shared their traditional celebrative meal of chicken and wine, and he dissolved the pills to provide a drinkable mixture. She drank it and was dead the next morning.

"The primary project in her life was to be independent," Mr. Delury said, and so she exercised that independence one last time on

1. Carey Goldberg, "A Meal, a Bitter Potion, a Major Test of a Law," *The New York Times*, 6 July 1995, p. A1.

July 4. He, for his part, hopes to write a book about death and, especially, about what he calls the "considerate hero" — an elderly person, for example, "who, facing serious illness, says, 'This money would be better spent elsewhere.'"

Here, in short is the story of a woman whose suffering evokes our sympathy — and also our fears, since we are uncertain what we would do in her circumstances. She is a woman whose "primary project" in life is to be independent, a project that surely calls for scrutiny, however great our sympathy for her. And her story, at least as told by her husband, invites us to consider whether we might be obligated to act heroically at some point in the future, ending our life in order to save resources that could be better spent on others. This is the sort of case that today makes news, and it turns our attention to two considerations — autonomy and suffering.

Christians have held that suicide is morally wrong because they have seen in it a contradiction of our nature as creatures, an unwillingness to receive life moment by moment from the hand of God without ever regarding it as simply "our" possession. We might think of ourselves as characters in a story of which God is the author. Dorothy L. Sayers ingeniously developed this analogy of artistic creation in *The Mind of the Maker.* Of the "work" produced by the artist Sayers writes:

> For the satisfaction of its will to life it depends utterly upon the sustained and perpetually renewed will to creation of its maker. The work can live and grow on the sole condition of the maker's untiring energy; to satisfy its will to die, he has only to stop working. In him it lives and moves and has its being, and it may say to him with literal truth, "Thou art my life, if thou withdraw, I die." If the unselfconscious creature could be moved to worship, its thanks and praise would be due, not so much for any incidents of its structure, but primarily for its being and identity.[2]

2. Dorothy L. Sayers, *The Mind of the Maker* (1941; New York: Harper & Row, 1979), pp. 141-42.

Characters in a story do, of course, have a real, if limited, freedom, and a good author will not simply compel them to do what is contrary to the nature he himself has given them. But at the same time characters do not determine the plot of their life's story, and it is a contradiction of their very being if they attempt to bring the story to its conclusion. We are dependent beings, and to think otherwise — to make independence our project, however sincerely — is to live a lie, to fly in the face of reality.

Thus, suicide as a rational project expresses a desire to be only free and not also finite — a desire to be more like Creator than creature. Of course, suicide may often result from depression or other emotional illness. In such cases it is not a rational undertaking, and we do not regard a person in such a state as a responsible agent. But there is no reason to deny that suicide can sometimes be undertaken by those who are not emotionally ill and who are responsible for their actions. Such suicide has about it a Promethean quality, a rejection of our status as creatures. Precisely because this is true, it is important to state that, contrary to what Christians have often believed, such rational suicide does not necessarily damn one. The suicide dies, so to speak, in the moment of sinning, without opportunity to repent. But then, so may I be killed instantly in a car accident while plotting revenge against an enemy of mine. God judges persons, not only individual deeds, and the moment in one's life when a sinful deed occurs does not determine one's fate. Even if I try to reject God and keep the light of his presence from my life, there is no guarantee that my creaturely action will be able to overcome his authorial ingenuity. Thus, the psalmist writes:

> Whither shall I go from thy Spirit?
>   Or whither shall I flee from thy presence?
> If I ascend to heaven, thou art there!
>   If I make my bed in Sheol, thou art there!
> If I take the wings of the morning
>   and dwell in the uttermost parts of the sea,
> even there thy hand shall lead me,

> and thy right hand shall hold me.
> If I say, "Let only darkness cover me,
> and the light about me be night,"
> even the darkness is not dark to thee,
> the night is bright as the day;
> for darkness is as light with thee.
>
> (Psalm 139:7-12)

Ultimate judgments about the person are not, therefore, ours to make, and we can condemn the act of suicide without claiming to render such a verdict.

What should be clear, though, is that Christians do not approach this issue by first thinking in terms of a "right to life" or a "right to die with dignity." That is to say, we do not start with the language of independence. Within the story of my life I have the relative freedom of a creature, but it is not simply "my" life to do with as I please. I am free to end it, of course, but not free to do so without risking something as important to my nature as freedom: namely, the sense of myself as one who always exists in relation to God.

For a society as much in love with individual autonomy as ours, this view may seem not only objectionable but also peculiar — and peculiarly religious. As such it will seem to be simply a quirky view held for private reasons by some people, but hardly a view that could have any public standing. We have taken autonomy so for granted, accepted it so much as the natural state of affairs, that we have lost our ability to question it or to see that — every bit as much as religion — it also presupposes a metaphysic and a view of human nature. When, in December 1964, an official of the Cuban revolutionary government named Augusto Martinez Sanchez committed suicide, Fidel Castro issued the following statement:

> We are deeply sorry for this event, although in accordance with elemental revolutionary principles, we believe this conduct by a rev-

> olutionary is unjustifiable and improper. . . . We believe that Comrade Martinez could not consciously have committed this act, since every revolutionary knows that he does not have the right to deprive his cause of a life that does not belong to him, and that he can only sacrifice against an enemy.[3]

To take an illustration much closer to home, one of my students once described her cousin's suicide and its continuing effects on his family by saying: "He didn't just take his own life; he took part of theirs too." Such examples from the political and familial spheres suggest how deceptive the language of autonomy may be. We do not have to approve Castro's political aims to recognize that the lives of fellow citizens may be bound together in such a way that all are aggrieved by the death of one. And the familial example reminds us that our identity is not an individual achievement but is socially formed from the very beginning. Christians simply extend that sociality of the self — believing that we exist always in relation to God, the author of our being who has authority over us. Not all will share that belief, of course, but there is nothing peculiar or unfamiliar about such an understanding of the self. It explains why we do not think our lives are our own, why we believe suicide to be morally wrong.

Caring (But Only Caring)

If my life is not simply my possession to dispose of as I see fit, as if the God-relation did not exist, the same is true of the lives of others. I have no authority to act as if I exercised lordship over another's life, and another has no authority to make me lord over his life and death. Hence, Christians should not request or cooperate in either assisted suicide or euthanasia.

3. Cited in Michael Walzer, *Obligations* (New York: Simon & Schuster, 1970), p. 172.

In a chapter written more than a quarter century ago, which remains to this day a classic discussion of the issues, Paul Ramsey sought to articulate an ethic of "(only) caring for the dying."[4] Such an ethic, he suggested, would reject two opposite extremes: refusing to acknowledge death by continuing the struggle against it when that struggle is useless, or aiming to hasten the coming of death. Neither of these can count as *care* for one of our fellow human beings; each is a form of abandonment. We should always try to care for the dying person, but we should *only* care. To try to do more by seizing either of the extremes is always to give something other or less than care. In the next chapter we will look closely at the problem of treatment refusals, of determining when it is permissible to cease the struggle against illness and death. In this chapter we concentrate on the temptation to hasten the coming of death.

Why might we be tempted to ask for or offer euthanasia? One of the reasons has already been suggested in the discussion of suicide: our commitment to autonomy or self-determination. I am tempted to believe that my life is my own to do with as I please — and tempted to believe that another's life is her own to do with as she pleases. A second reason, equally powerful and tempting, is our desire to bring relief to those who suffer greatly. The argument for euthanasia rests chiefly on these two points, taken either singly or together.

Usually in the public debates of our society these two reasons are presented as a package, as if they must be taken together. For the present, that is, advocates of euthanasia (or, as it is sometimes termed, assistance in dying) have tended to argue that it should be permitted only if the person euthanized was suffering greatly *and* (while competent) had requested such assistance. One suspects, however, that this is a purely strategic maneuver aimed at keeping the argument for the time being a relatively narrow one. Whatever our judgment of motives,

4. See chapter three of Paul Ramsey, *The Patient as Person* (New Haven and London: Yale University Press, 1970).

however, the fact is that, simply as a matter of logic, the two prongs of the argument will gradually become independent of each other.[5]

If self-determination is truly so significant that we have a right to help in ending our life, then how can we insist that such help may rightly be offered only to those who are suffering greatly? Others who are not suffering may still find life meaningless, the game not worth the candle. They too are autonomous, and, if autonomy is as important as the argument claims it is, then their autonomous requests for euthanasia should also be honored, even if they are not suffering greatly. Similarly, if the suffering of others makes so powerful a claim upon us that we should kill them to bring it to an end, it is hard to believe that we ought to restrict such merciful relief only to those who are self-determining, who are competent to request it. Surely, fully autonomous people are not the only human beings who can suffer greatly. Thus, from both directions, from each prong of the argument, there will be pressure to expand the class of candidates for euthanasia. Those who suffer greatly but cannot request relief and those who request help even though their physical pain is not great will begin to seem more suitable candidates. That is, in fact, the logic of the argument currently being played out in our public policy disputes.

Whatever the outcome of arguments in the public sphere, of course, Christians must form their own views. If assisted suicide and euthanasia continue to become increasingly accepted, we must dissent from that attitude. For neither prong of the argument in support of euthanasia grows out of Christian conviction. The autonomy argument has already been taken up in this chapter's discussion of suicide, where we noted its defective, individualistic understanding of the human person, but that does not exhaust the argument's difficulties. For Christians, each person's life is a divine gift and trust, taken up into God's

5. This argument has been pressed forcefully by Daniel Callahan in a number of places. See especially chapter three of *The Troubled Dream of Life* (New York: Simon & Schuster, 1993).

own eternal life in Jesus, to be guarded and respected in others and in ourself. Because, however, we are inclined to overemphasize our freedom and forget the limits of our finite condition, inclined to forget that life comes to us as a gift, death becomes the great reminder of those limits. In *The Death of Ivan Ilych* Tolstoy powerfully captures Ilych's surprise that he — and not just others — should come up against this limit, that his own existence should be included in the truth that all men are mortal. But try as we may to forget it, death *is* the starkest reminder of our limits. It is, therefore, a peculiar moment at which to attempt to seize ultimate control of our life and pretend that we are independent self-creators. That is simply one last way of living a lie.

Moreover, euthanasia is not simply an extension of personal autonomy; it is not simply "nonintervention" in another person's private choice. On the contrary, because it requires the participation of at least one other person, it becomes a communal act involving the larger society and giving its approval to an act of abandonment. If it becomes a permissible and acceptable practice, have our freedom and independence been enhanced? In one sense no doubt they have, since we are given a new option, the option to die (with help) when we wish. But it is also true that the pressure will build to exercise this option, to be the "considerate hero" who does not stay alive too long using resources that might better be spent elsewhere. If and when euthanasia receives social approval, what looks like more freedom is likely to turn out to be less. In short, there are good reasons not to acquiesce in the autonomy argument.

Christians are, I suspect, more likely to be drawn to the argument that describes euthanasia as compassionate relief of suffering. And, to be sure, we all know the fear of suffering and the frustration of being unable to relieve it fully in those whom we love. The principle that governs Christian compassion, however, is not "minimize suffering." It is "maximize care." Were our goal only to minimize suffering, no doubt we could sometimes achieve it most effectively by eliminating *sufferers.* But then we refuse to understand suffering as a significant part of human life that can have meaning or purpose. We should not, of course, pretend that suffering in itself is a good thing, nor should we put forward claims about the benefits others can reap from their suffering. Jesus in Gethsemane — who shrinks from the suffering to come but accepts it as part of his calling and obedience — should be our model here. The suffering that comes is an evil, but the God who in Jesus has not abandoned us in that suffering can bring good from it for us as for Jesus. We are called simply to live out our personal histories — the stories of which God is author — as faithfully as we can.

Our task, therefore, is not to abandon those who suffer but to "maximize care" for them as they live out their own life's story. We ought "always to care, never to kill."[6] And it has, in fact, been precisely our deep commitment not to abandon those who suffer that has, in large measure, been a powerful motive force in the development of modern medicine. Our continuing task is not to eliminate sufferers but to find better ways of dealing with their suffering. If we cannot always fully relieve the suffering, when we cannot relieve it, we must remember that even God does not really "solve" or take away the problem of suffering; rather, God himself lives that problem and bears it. His way is steadfast love *through* suffering, and it is the mystery of God's own being and power that this truly proves to be the way to maximize care for all who suffer.

Christian physicians will have their own special reasons as physicians to refuse involvement in the practice of euthanasia. They have found in medicine what we all seek — work that can truly be understood as a calling. Knowing themselves to be whole only by the grace of God, they can seek to impart a small measure of wholeness — healing and health — to their patients. For several millennia the tradition of Western medicine has followed the Hippocratic Oath's injunction not to "give a deadly drug to anybody if asked for it, nor . . . make a sug-

6. See "Always to Care, Never to Kill: A Declaration on Euthanasia," *First Things*, February 1992, pp. 45-47.

gestion to this effect." But physicians swim daily in a sea of human suffering, and they may be especially tempted, when all else has failed, to eliminate suffering by eliminating the sufferer. Christian physicians, who understand their life as response to the grace God has extended to them in their perishing condition, should, in William F. May's words, be set free "from the need to avoid ties to the perishing."[7]

It should therefore become part of their calling — at least in this time and place — to bear witness to all of us that suffering and death "are *real* but not *ultimate;* [and that] they do not speak the last word about the human condition."[8] Good physicians will know the limits of their art, and they can help us avoid the notion that there is any ultimate "technological fix" for the fundamental human problems of suffering and death. This means, we should note, that for physicians as well as the rest of us there are limits to what we should do in our attempts to relieve suffering. A willingness to discern such limits as best we can — and, having discerned them, to act in accord with them — is deeply embedded in the Christian understanding of the moral life. Understanding compassion and care in this way, we seek to learn to stand with and beside those who suffer — with them as an equal, not as a lord over life and death, but determined not to abandon them as they live out their personal histories up against that limit of death which we all share. For us, therefore, the governing imperative should be not "minimize suffering," but "maximize care."

7. William F. May, *The Physician's Covenant* (Philadelphia: Westminster Press, 1983), p. 127.

8. May, *The Physician's Covenant,* p. 127.

CHAPTER SEVEN

Refusing Treatment

This chapter might be described simply as an extended reflection on David Smith's insightful metaphor of the "good host."

> A couple invite friends to dinner. Food and drink are pleasant; the conversation bubbles. The good host is hospitable and courteous to his guest, no matter what his shifts in mood. But there comes a time when the party "winds down" — a time to acknowledge that the evening is over. At that point, not easily determined by clock, conversation or basal metabolism, the good host does not press his guest to stay but lets him go. Indeed he may have to signal that it is acceptable to leave. A good host will never be sure of his timing and will never kick out his guest. His jurisdiction over the guest is limited to taking care and permitting departure.[1]

The good host recognizes certain boundaries. He neither kicks out his guest nor presses him to stay when the time has come for the party to end. These boundaries correspond to the two opposite extremes which Paul Ramsey suggested we had to avoid in caring for the dying.

1. David H. Smith, *Health and Medicine in the Anglican Tradition* (New York: Crossroad, 1986), p. 52.

On the one hand, we ought not choose death or aim at death. But on the other hand, neither should we act as if continued life were the only, or even the highest, good. It is not a god, but a gift of God. Thus, we should neither aim at death nor continue the struggle against it when its time has come. "Allowing to die" is permitted; killing is not. Within these limits lies the sphere of our freedom.

Aim and Result

We need therefore to explore that sphere of limited freedom, to clarify as best we can when it is permissible *not* to treat. Before we can approach that question directly, however, we need to unpack the difference between the *intention* or *aim* of an action and the *result* of an action. If I commit suicide (and am of sound mind), I intend to die. I aim at my death or choose death. But, of course, there might be occasions when, if I refuse a certain treatment, I will also die. Both the act of suicide and the treatment refusal may *result* in my death. Are they therefore morally equivalent? Is treatment refusal the same as the forbidden suicide? Although they could sometimes be morally equivalent — I could refuse treatment *so that* I will die — they need not be. To see why we must think about the *aim* and the *result* of an action. Some examples, first of all from outside the sphere of medicine, will help.

A soldier may charge the enemy, knowing that he faces almost certain death in so doing. He does not thereby commit suicide. He does not *choose* to die, even though he foresees that death is the likely, perhaps almost certain, *result* of his action. We might even say of the soldier that his mission was a "suicidal" one, meaning that his death would almost certainly result from it. But we wouldn't call what he does an "act of suicide," for he doesn't intend his death or aim at it. Dying is not part of his plan of action, just its very likely result.

A distinction like this has been important for Christians in very different circumstances. Christians generally forbade suicide, but they honored their martyrs. Why? How can they honor someone whose own choice so certainly results in her death? They honor the martyr because she does not *aim* at her death. She aims to be faithful to God, foreseeing as a likely result the loss of her life. Forbidding suicide and honoring martyrs, Christians recognized life as a real but not ultimate good — a great good, but not the highest good.

We can begin to approach the problem of treatment refusals more directly if we consider a medical example. A believing Jehovah's Witness, having been injured in an accident, might need a blood transfusion and might, as a matter of religious principle, refuse it. The *result* of that choice could be his death, but he is not choosing or intending to die. He is choosing to be faithful to God as he understands such faithfulness. If it transpired that he survived without the transfusion, he would not feel that his aims had been frustrated or his plan of action had failed; for he was not aiming to die. By contrast, to take a bizarre possibility, suppose others were trying to force the transfusion upon him and, in order to stop that from happening, he grabbed a gun and shot himself. That would be an act of suicide. He would be aiming at his death, embracing it as a part of his plan of action. But that is just what he does not do in refusing the transfusion.

One more example is needed to unpack the full implications of our distinction between an act's aim and its result. A patient in the last stages of terminal illness, who is suffering greatly, may request and receive increasingly large doses of morphine to control his pain. We know that increasingly large doses of a narcotic drug may bring death more quickly by suppressing respiration. That is one possible *result* of this treatment, and of course one could aim at that result by giving a dosage large enough to cause death. But a carefully calibrated increase in the amount of medication is aimed at controlling pain, not at bringing a quicker death. The result may, in fact, turn out to be a slightly quicker death than would otherwise have happened, though that is hard to know. The aim, however, is to provide the best care possible in these difficult circumstances. Neither the patient who requests the morphine nor the doctor who authorizes it is necessarily choosing death in so doing.

This distinction between an act's aim and its result is crucial to bear in mind when we consider decisions to refuse or withdraw treatment. The *result* of such decisions may be that death comes more quickly than it might have. Nevertheless, the fact that we ought not aim at death for ourself or another does not mean that we must always do everything possible to oppose it. Life is not our god, but a gift of God; death is a great evil, but not the ultimate evil. There may come a time, then, when it is proper to acknowledge death and cease to oppose it. Our aim in such circumstances is to care for the dying person as best we can — which now, we judge, means withdrawing rather than imposing treatment.

Guidelines

How can we recognize that time when it has come? There are answers to this question, but they are not answers that admit of mathematical precision. The point at which the party has wound down, we recall, is "not easily determined by clock, conversation, or basal metabolism," and the "good host will never be sure of his timing." He will never kick out his guest but will seek only to take care of the guest and permit departure. We are, in short, seeking guidance that will help us know when refusing treatment is not aiming at death, when it is not the forbidden suicide or euthanasia.

Traditionally, the language of *ordinary* and *extraordinary* (or *heroic*) was used to mark this point. "Ordinary" care was obligatory and not to be refused for oneself or for another. "Extraordinary" care was optional; one could accept or decline it depending on one's own life plans. This language has often proved to be more puzzling than clarifying. Ordinary treatment was sometimes thought to be treatment that is common or simple as opposed to treatment that is rare or technologically complicated. But those are not moral distinctions, and, in fact, a treatment that was both rare and complicated could be "ordinary" in the moral sense. If this language does not serve us well, though, its point is still one that is crucial. We need to know what makes a treatment the sort we can refuse, what makes a treatment what used to be called "extraordinary." Remembering that mathematical precision is not possible here, we can delineate two criteria for treatments that may be refused. Either taken by itself gives sufficient moral reason for refusing or withdrawing treatment.

First, a treatment may be refused if it is useless. Usefulness is, of course, relative to the patient's condition. Many treatments may be helpful, even lifesaving, for some patients but of little use to others. Moreover, a treatment once helpful might cease to be when the patient's disease has progressed to a new stage. This criterion is especially important when a person is in the last stages of dying. (We should note, however, that to be irretrievably dying is not the same as to be terminally ill. One can be terminally ill but still expected to live for months or even years.) For the patient who is irretrievably dying, few if any treatments can really be useful. Continued attempts to *cure* such a patient may well get in the way of the effort to *care* for this person as best we can. In any case, no one is obligated to pursue treatments that are not expected to be helpful, and to refuse such treatment is exactly that: the refusal of a treatment, not the rejection of the gift of life. It is not killing but "allowing to die."

Although this first guideline is clear, it can easily be abused when applied. Consider, for example, the case of a baby born with Down's syndrome and with an obstruction of the esophagus. In order for the infant to be fed, surgery is required to remove the obstruction, and we would surely do such surgery for an otherwise "normal" infant. Yet, there have been cases in which parents — upheld by courts — have refused treatment for such an infant. Has that child simply been "allowed to die"? Or have those responsible for its care in fact taken aim at its life? If we were the parents who had decided to refuse surgery, how might we formulate our plan of action? The most accurate way to describe our plan would be to say: We will refuse the surgery *so that* the child will die. Such a plan is not properly described simply as "allowing to die." The aim is not to care for the child's life as best we can but to end it, and we deceive ourselves if we think otherwise.

The language of "personhood" has encouraged such self-deception. It has encouraged us to think that we do no wrong when we deliberately "let die" (that is, refuse treatment in order to aim at the death of) those human beings who lack some of the distinctively human cognitive capacities. To be sure, useless treatments may rightly be refused, but we need to make certain that we ask of possible treatments: Will they benefit the life this patient has? That is quite different from asking, Is this patient's life a beneficial one, a life worth living? If we find ourselves asking the second question, then, quite clearly, we are taking aim not simply at a treatment but at a life. That should not be called "allowing to die."

There is a second reason — in addition to uselessness — on the basis of which treatments may rightly be refused. Treatments that are useful and perhaps even lifesaving may sometimes be *excessively burdensome*. Because life is not our god, we need not accept all burdens — no matter how great — in order to stay alive. We need to recognize clearly what this means. It means that we might rightly refuse even useful treatment that would prolong our life for a significant period of time if that treatment really does carry with it significant burdens.

To reject or withdraw treatment because of its burdens is still a refusal of treatment, not of life. From among the various lives still available to a suffering patient — some longer than others; some filled with more burdens than others — we choose one life in particular. Just as the soldier going on a suicidal mission does not choose to die but, rather, to live in a certain way, recognizing that to live in this way may mean not to live as long, so also the patient refusing an excessively burdensome treatment still chooses life — one particular life from among the several still available.

This criterion can also be abused. It guides us in our decision making, but it does not make decisions for us. It trains us to distinguish between the burdens of treatment and the burden of life, a distinction we are often tempted to pass by. The cases of Karen Quinlan and Nancy Cruzan, each well known, illustrate the complexities of this temptation. Both young women became permanently unconscious in a persistent vegetative state and were fed through a tube to sustain their life. Both had suffered irreversible loss of the higher brain functions that make consciousness and self-awareness possible. We might easily be tempted to think that their lives were useless, that it was a burden even to have such a life. That is an understandable thought, for none of us would choose such a life for ourselves if given alternatives. But if we act on such a thought and withdraw the feeding tube, the burden at which we are taking aim is not treatment but life itself.

In the case of certain patients especially — the severely demented, the permanently unconscious, the profoundly retarded or disabled — we may sometimes be tempted by the thought that their lives are a burden not worth continuing. But we need honestly to recognize that if we reject treatment for that reason, because the life itself is a burden, then we are rejecting not just treatment but life. We have ceased to care for that person as best we can in the time and place he has been given.

To summarize: Treatment may be refused or withdrawn when it is either useless or excessively burdensome. In either of those instances, refusal of treatment is not the forbidden suicide or euthanasia. Although the distinctions I have drawn here would be difficult to establish in a court of law, where the aims of agents are not always apparent, they are necessary for the moral life. If we can honestly describe a possible treatment as either useless or excessively burdensome, then in rejecting that treatment we can still choose life. But if the treatment itself carries no excessive burden (even though, of course, the patient's life itself may be burdensome), and if the treatment will benefit the life the patient has (even though, given alternatives, we would not desire that life), we ought to choose life both for ourselves and for others.

Truth-telling

In the presence of dying patients we all stammer, uncertain what to say. Quite often we say less than the truth — sometimes deliberately, sometimes simply because we cannot rise to the momentousness of the occasion. But patients cannot make thoughtful decisions about their treatments if they do not know the truth of their condition; hence, we must learn to speak the truth with them, to them, and for them. "Truth" is not a simple concept, however. One of the best brief descriptions of its complexities has been provided by "Miss Manners."

> You have heard . . . of "the truth, the whole truth and nothing but the truth." . . .
>
> The whole truth is the moral answer to such questions as "Did you take a cookie after I told you not to?" Such an answer might be, "No, but I ate the cooking chocolate." One must always speak the whole truth when morality is at stake.
>
> Nothing but the truth is more useful on social occasions. For example: The question, "How do I look?" may be answered with, "To me, you're always beautiful," even though the whole truth would require adding, "but that dress makes you look like a truck."
>
> The truth is the most complex concept of all. It means getting to the truth of the situation, rather than the crude literal surface truth. To answer the question, "Would you like to see some pictures of my grandchildren?" with the direct literal truth, "No! Anything but that!" would be cruel. But is that the real question? The real question, if one has any sensitivity to humanity is, "Would you be kind enough to let me share some of my sentiments and reas-

> sure me that they are important and worthwhile?" to which a decent person can only answer, "I'd love to."[2]

Seriously ill and dying patients are not always in need of what Miss Manners calls the whole truth, and we do not necessarily lie if — at least for a time — we disclose less than we know. We and those for whom we care may be well served by what William F. May calls "the language of indirection," which "treats death decorously as a sacred event," avoiding the "gabby bluntness" that fails to take seriously the awesomeness of a person's dying.

> A man who knew that he had cancer once said to his middle-aged son, a writer on the subject of death and dying, "Go easy, Don." The man knew he had cancer. But at the same time, he wanted to establish the distance he wished to maintain between himself, his son, and the imminent event of his death. He did not want his son to favor him with seminar-length discussions on the subject. Only a fool would not have respected this request.[3]

There are other reasons as well why "indirection" is often in order. Some truths cannot be received at just any moment. (We cannot explain the "facts of life" to a two-year-old.) We cannot bring the truth of her condition to a dying woman simply by confronting her with it. She must herself be ready to hear. Tolstoy's Ilych does not come to know the truth of his condition by recalling that all men are mortal. True though that statement is, it is not known in the same way he comes to know that *he* is dying. As long as we think of ourselves as having taken the first step on a road that leads ultimately toward "the whole truth," we do not necessarily lie or deceive when we disclose less than we know.

2. The *Cleveland Plain Dealer,* 21 June 1984, p. 5-E.

3. William F. May, *The Physician's Covenant* (Philadelphia: Westminster Press, 1983), p. 164.

"Nothing but the truth" is what Miss Manners prescribes for social occasions. Helmut Thielicke, for example, recalls

> a sexton at whose church theological students frequently did the preaching. He always had three stock answers when they asked with anxious curiosity how they had done. If they had done well he would reply, "The Lord has been gracious"; if moderately well, "The text was difficult"; and if badly, "The hymns were well chosen."[4]

Such delicacy, Thielicke notes, is really more than a useful device for social occasions; for it expresses a deep respect for the other person. In so doing it points beyond itself toward "the truth."

This is less a matter of any particular form of words than it is of being "in the truth." If we are in the truth, we will not be quick to suggest to any patient that his condition is hopeless. Such a suggestion runs the risk of verifying itself by destroying the patient's will to live. Hope, after all, is the virtue that is needed precisely when all seems lost. Its internal dynamic, therefore, presses us toward the transcendent, toward a power greater than our own. If we are in the truth, we will not deny that such power may help when we cannot. But neither should we fail to reckon with what seems to be happening, for we are not likely to be given another opportunity. To be "in the truth" is to be forced to face our own uncertainty in the face of death. From that perspective we are not securely in possession of truth that we may impart to the dying person. On the contrary, our own security has also been shaken, and together we must seek to learn "the truth." We should not simply impose "the whole truth"; we should learn the deep respect that underlies "nothing but the truth," and we should seek to bring the person in our care to that moment in which he can know "the truth" of his condition — the moment of death.

4. Helmut Thielicke, *Theological Ethics,* volume 1: *Foundations* (Philadelphia: Fortress Press, 1966), p. 549.

CHAPTER EIGHT

Who Decides?

Our focus in the last chapter was on the substance of treatment decisions — *what* we should do and what criteria governed those choices. Sometimes, however, we may wonder or argue about *who* should decide. We might, of course, simply ask our doctors to decide what is best, and there was a time not that long ago when doctors' recommendations about treatment were seldom disputed. The past few decades, however, have seen a strong rejection of medical paternalism and an increasing emphasis upon patient self-determination. The presumption now is that the patient decides what course of treatment shall be pursued.

Having taken the measure of the language of self-determination in Chapter 6, we should be wary of depicting the possibilities for treatment decisions in this way. Even fully autonomous patients, if there are such, have no absolute right to decide upon their course of treatment. If they did, physicians would simply be technicians, putting their skills (for a price) at the service of our desires. However tempted we might be by that picture, however often medical hubris or paternalism may push us toward it, we should not really want it.[1] A *patient* is something

1. We should not want it because it misses the *moral* truth of medicine. In this country, however, the *law* has moved increasingly in the direction of patient auton-

different from a *client,* and a physician is not an automobile mechanic. When he examines, handles, and even cuts upon our body, the doctor lays hold upon our person. As patients, therefore, we quite rightly should be involved in deliberations about the course of our treatment; for our person is involved. But so is the doctor's person involved as he commits himself to care for us. We should not want it any other way.

There may be times, of course, when no agreement is reached between doctor and patient, and the best we can do is withdraw from the mutual bond we have forged. Lacking agreement on the substance of the matter, we take refuge in a procedural solution that leaves both parties free to turn elsewhere. Patients need not submit to doctors' recommendations; doctors need not practice what they consider bad medicine simply because patients want it. In a society such as ours — where substantive agreement is sometimes so lacking that a commitment to "fair procedures" is almost all we share — Christian physicians often find themselves in difficult situations. Not wishing to abandon patients who disagree with their judgments about the best course of treatment, they may feel drawn or even compelled to practice what they regard as bad medicine (which is a moral, not just a technical, category). Moreover, there is no guarantee that the medical profession itself will support Christian principles in the practice of medicine. That is already the case with respect to abortion and prenatal screening, and it may increasingly be true of assisted suicide and euthanasia. Christian care-giving institutions, such as hospices, will face similar difficulties if assisted suicide and euthanasia become legally permissible. To be sure, it is one thing to acquiesce in a patient's decision to pursue a course of treatment when the physician simply thinks there are better and wiser courses; it is another to do moral evil in the name of not abandoning one's patient. Christian physicians in our society will probably have to take increasing care to make clear to their patients

omy. Here, though, I am reflecting upon what our moral judgment as Christians ought to be when we consider the relation of patient and physician.

from the outset their own understanding of good medical practice and the limits to which they adhere.

Incompetent Patients

Even if patients' wishes are not entirely determinative, they may quite rightly want to be involved in deliberations about their course of treatment. (Of course, that involvement may also take the form of saying what they do *not* want to know or intentionally leaving decisions to others, and we should also respect that form of involvement. Personal involvement is not demonstrated only through a concern for mastery and control.) Sometimes, however, patients are unable to participate in deliberations about treatment. This may happen for a short time due to the trauma of injury, but more difficult are cases of patients who will be incompetent to help make decisions for the entire course of their treatment — infants and young children, the severely demented, the retarded, the permanently unconscious. In these cases we are forced to ask in earnest, Who decides?

In recent years the professions of both law and medicine have recommended advance directives — either a living will or a health care power of attorney — as the best way to answer this question. In essence, it attempts to circumvent the question by having the patient, while still competent, determine and state how he wants to be treated if and when he should become incompetent.

Not all patients will have executed an advance directive, however, and, in the very nature of the case, some patients will never be able to do so (because, for example, they are infants, or have been retarded from birth). In those circumstances some have turned to what is called a "substituted judgment" standard, and, in fact, the law sometimes compels us to turn in that direction. According to this approach, we should ask what a patient would have wanted if he were able to tell us. There are, of course, different ways to try to answer such a question.

We might assume that he would want what any "reasonable person" would want in his circumstances. But then we may discover that we have no agreed-upon standard of "reasonableness." The blood transfusion that seems reasonable to me will look quite different to the faithful Jehovah's Witness. The lengthy round of chemotherapy that seems choiceworthy to you may look quite undesirable to me.[2]

Taken seriously, therefore, the substituted judgment standard may direct our attention away from the hypothetical reasonable person to the actual person who is the patient. Sometimes we know (from family or friends, for example) a good bit about what this person might have wanted. Sometimes the evidence may be sketchier — when, for example, he once opined in a late-night conversation with friends, "I'd never want to be kept alive on any machines." And sometimes, of course, the patient will be one who has no "track record" of past opinions or decisions — for example, a newborn, or a person who has been profoundly retarded from birth. For such patients substituted judgment seems inappropriate, although courts have sometimes applied it. Thus, for example, in the case of Joseph Saikewicz, who had been retarded from birth and at age sixty-seven suffered from a form of leukemia for which chemotherapy was a possible treatment, the Supreme Court of Massachusetts, attempting to apply a substituted judg-

2. The difficulty of determining what is "reasonable" is even more troubling when the patient is a child for whom parents ordinarily make decisions. For example, Christian Scientists, who often turn to their own practitioners for faith healing rather than to standard medical practice, have sometimes been prosecuted for child neglect when they have not sought standard care for their seriously ill children. These are hard cases. No society can survive without some minimal agreement about what constitutes "reasonable" care for one's child. Indeed, the very existence of child abuse and neglect laws indicates that, and we would not permit a religious sect to revive the ancient practice of sacrificing the first-born son to God. That would fall beyond the bounds of "reasonableness" for us. Nevertheless, we ought to be reluctant to narrow too quickly our understanding of what is reasonable in approaches to medical care. Impressive as the healing powers of the medical profession are, they are not the only healing powers. To suppose that they are would, for Christians, truly be "unreasonable."

ment standard, held that "the decision in cases such as this should be that which would be made by the incompetent person, if that person were competent, but taking into account the present and future incompetency of the individual as one of the factors which would necessarily enter into the decision-making process of the competent person."[3] The thought of an utterly hypothetical Saikewicz — who is not the real Joseph Saikewicz at all — being given one moment of lucid rationality in which to decide whether, as the person he actually is, he would want chemotherapy is itself a *reductio ad absurdum* of the attempt to apply a substituted judgment standard in such cases.

It is only our nearly idolatrous attachment to the language of autonomy that drives us to such lengths, of course, and where the law will permit it we should not hesitate to turn from substituted judgment to an attempt simply to assess what is in the patient's *best interests.* To be sure, there is no guarantee that this language will not also lead us astray. For patients with severely diminished capacities, we may too easily be influenced by the fact that we would not desire such a life for ourselves. Instead of asking, "Is his life a benefit to him?" we need to learn to ask, "What, if anything, can we do that will benefit the life he has?" Our task is not to judge the worth of this person's life relative to other possible or actual lives. Our task is to care for the life he has as best we can. Properly applied, a best interests standard can free us from the often futile quest for what he would have wanted. It can free our energies and direct them toward the right question: Given the person he is now and has become, how can we best nourish and care for the life he has?

3. Supreme Judicial Court of Massachusetts, 1977. 373 Mass. 728, 370 N.E.2d.417. The case is reprinted in a variety of sources. See, for example, Thomas A. Shannon and Jo Ann Manfra, eds., *Law and Bioethics: Texts with Commentary on Major U.S. Court Decisions* (Ramsey, NJ: Paulist Press, 1982), pp. 173-92.

Advance Directives

Obviously, some of these difficulties, for some patients, can be avoided if they have expressed their treatment preferences in advance — if, that is, they have, while competent, formally stated how they wish to be treated if a day comes when they are incompetent and unable to participate in decision making. An advance directive is an attempt to extend our autonomy into a future time when we are no longer autonomous. As such, it is a product of the emphasis upon self-determination within our society over the last several decades, and it even has about it an illusory quality that attempts to give privileged status to one moment of independence in the course of an entire life that begins in dependence and, often, ends in dependence. Therefore, if they are not used with care, advance directives give rise to a kind of metaphysical self-deception.

Two forms of advance directives have been developed. What is called a "living will" was first given legal standing by the state of California in 1976. In enacting a living will I attempt to describe in advance the possible medical conditions that might overtake me in the future, and I attempt also to stipulate how I would want to be treated (or not treated) under those conditions. Developed at a time when the chief concern was the heavy hand of medical paternalism, the living will has often been conceived as an instrument for refusing treatment, for getting rid of that heavy hand. In principle, however, there is no reason why one could not use such an instrument to express a desire *for* treatment, even for all possible treatments. And, of course, the laws of any state may set limits on the treatments one can refuse or require. By contrast, a health care power of attorney attempts to say less about the future. Eschewing the attempt to predict possible medical conditions or treatments, it simply designates a proxy — one who will be authorized to participate in decision making on my behalf if I become unable to do so.[4]

4. There are also mixed or combined advance directives in which one both names a proxy and provides that proxy with information about one's treatment pref-

There is, I think, no single "Christian" position on advance directives, but, in my judgment, we would not be wise to make use of the living will. After the U.S. Supreme Court issued its decision in the Nancy Cruzan case (in 1989), it was reported that the Society for the Right to Die received over 100,000 requests for information about living wills in less than a month. This testifies to an enormous sense of dis-ease within our culture. Even though the normal human biography begins and ends in dependence, we deeply desire independence. That is understandable, of course, and in itself quite appropriate. But it ceases to be appropriate when it invites and encourages us to live a lie. When we attempt so definitively to extend our autonomous choices into that period of life when we are no longer self-determining, we come very close to such self-deception. In part, we deceive ourselves into supposing that we can actually anticipate with precision future medical conditions and possible treatments — a supposition that physicians themselves regularly resisted until lingering guilt over medical paternalism and the onslaught of malpractice litigation led them to acquiesce more readily in anything that appeared to be a patient's decision. More important, however, is the deception that cuts more deeply into our sense of self and encourages us to approach even the grave in a spirit of mastery and control.

Moreover, a living will lets others off the hook too easily. Patients who are unable to make decisions for themselves because, for example, they are severely demented or permanently unconscious have, in a sense, become "strangers" to the rest of us. We see in them what we may one day be, they make us uneasy, and we react with ambivalence. No matter how devoted our care, our uneasiness with a loved one who has become a stranger to us may prompt us to do less than we ought to sustain her life. It is important, therefore, to structure the medical decision-making situation in such a way that conversation is forced

---

erences. I do not treat this as a separate alternative because, as far as I can tell, it is best described as a slightly different kind of living will.

among the doctor, other caregivers, the patient's family, and the pastor or priest. Advance directives, often with the force of legal recognition standing behind them, are designed to eliminate the need for such extended conversation. That is part of their problem, for they free us from the need to deal with the ambivalence we feel in caring for a loved one who has now become a burdensome stranger.

I realize, of course, that freeing loved ones from such burdens is supposed to be one of the benefits of a living will, but Christians ought to be wary of such language.[5] For to burden one another is, in large measure, what it means to belong to a family — and to the new family into which we are brought in baptism. Families would not have the significance they do for us if they did not, in fact, give us a claim upon each other. At least in this sphere of life we do not come together as autonomous individuals freely contracting with each other. We simply find ourselves thrown together and asked to share the burdens of life while learning to care for one another. Often, of course, we will resent such claims on our time and energy. Indeed, learning not to resent them is likely to be the work of a lifetime. If we decline to learn the lesson, however, we cease to live in the kind of community that deserves to be called a family, and we are ill prepared to live in the community for which God has redeemed us — a community in which no one stands on the basis of his rights, and all live by that shared love Christians have called charity.

I think, therefore, that we ought to prefer the health care power of attorney to the living will.[6] It too, of course, reaches out into a future beyond the limits of our competence, but it does so in a way that recog-

5. In this paragraph and the preceding one I draw upon the language of my short piece titled "I Want to Burden My Loved Ones," *First Things*, October 1991, pp. 12-14.

6. There may, of course, be some people — especially quite elderly people — for whom no close relatives remain as possible proxies. They may understandably feel driven to enact a living will. One wonders, however, whether the church could not provide better alternatives, whether a proxy within the Body of Christ could not be found.

nizes and affirms dependence. It anticipates and accepts that others will have to bear some burdens for us as we may for them. To medical caregivers it says simply: "Here is a person upon whom I have often been dependent for love and care in the past. Now, when I can no longer participate in decisions about my medical care, I am content to continue to be dependent upon his love and care. Talk with him about what is best for me." In the cultural circumstances in which we find ourselves, I do not think Christians can do better than this.

CHAPTER NINE

Gifts of the Body: Organ Donation

Almost 80,000 Americans are currently on waiting lists, hoping to receive a donated organ. Many of these — especially those awaiting a heart or liver transplant — face situations that are immediately life-threatening, and they will die if a suitable organ for transplant is not found quickly. (In 2001 almost 24,000 patients in the United States received a transplant, though approximately 6,000 patients died while still awaiting a transplant.)[1] These donated organs — gifts of the body coming sometimes from those who have died, sometimes from those still living — are often truly life-saving. While it is true that patients undergoing organ transplantation may endure a great deal simply in order to stay alive, as techniques improve, so do survival rates. Thus, for example, five-year survival rates for those who have received a kidney transplant from a living donor approach 90 percent. (Graft survival rates are not actually this high, since some patients may have a second transplant if a first fails, and some may return to dialysis.) The rates are, of course, considerably lower for some other, more medically complicated, kinds of transplantation — not quite 50 percent for recipients of heart-lung transplants.

1. 2002 Annual Report of the U.S. Organ Procurement and Transplantation Network. See: www.ustransplant.org/annual_reports/ar02/highlights/ar02_chapter_one.htm. Accessed 6 February 2004.

At the same time that such gifts of the body help some to survive longer than they otherwise could, about 10,000 Americans die each year in circumstances — often because of accidents causing severe head injuries — that make them potential organ donors. After they have been declared dead because of the complete cessation of brain activity, but while heart and lung activity in the corpse is being artificially maintained to prevent organ deterioration, both tissues and organs — cornea, heart, lung, kidney, liver — can be taken for transplantation.

A 1995 study indicated that, despite the seeming need for organs, only about 40 percent of the 10,000 actually become donors. The percentage is somewhat higher now, but, even so, it remains true that organs are never taken for transplant from many who might make suitable donors. In most cases the reason is that the dead person's family refuses a request to take the organs for transplant.[2] *If* such refusal is a thoughtless act, one might argue that we should look for new ways to persuade families to consent to organ donation after death of a loved one. Thus, the death of Mickey Mantle after a liver transplant was seized as the occasion for encouraging organ donation. Many people who might never attend a talk on transplantation or read a magazine article on the subject may be influenced by hearing Bob Costas talk about the importance of this effort to remember Mantle. *If* such refusal is not just thoughtless but morally wrong, one might argue that we should override it — perhaps by authorizing medical professionals routinely to salvage, without prior request for approval, cadaver organs for transplant. And an increasing number of voices seem prepared to argue for that. *If* such refusal is motivated by concerns that are selfish or, at least, self-regarding, one might argue that we should fight fire with fire by offering to compensate the family of the deceased for donated organs — appealing to their self-regarding impulses in order to achieve desirable social aims. And such arguments have also found

2. Gina Kolata, "Families Are Barriers to Many Organ Donations, Study Finds," *The New York Times*, 7 July 1995, p. A9.

adherents within our society. On the other hand, *if* weighty — albeit often unarticulated — reasons may underlie such refusals, we ought to look with a critical eye upon the increasing social pressure to encourage organ donation and transplantation. These are not just questions of public policy. They are also questions that pit our deep-seated hunger to live longer and our fear of death against equally deep-seated notions of the sacredness of human life *in the body*. I will consider here the path we have traveled in approving and encouraging organ donation. Without rejecting that path or denying the good that comes from life-saving gifts of the body, we need to reflect upon what Christians in particular should think about a technique that promises much good but, also, raises profound questions about the meaning of our humanity. Our public rhetoric almost always lauds organ donation, the better to stimulate potential donors, but we need also to learn circumspection and caution in the use of such rhetoric.

Giving and Receiving Organs

In the late 1960s the Uniform Anatomical Gift Act was passed into law in every state in this country. It allows individuals, while still living, to authorize the donation of any parts of their body after death. If the deceased person had not authorized such donation but had not specifically prohibited it, specified family members are permitted to give authorization. The National Organ Transplantation Act, passed by Congress in 1984, established a national registry and donor-recipient matching system while also prohibiting the *sale* of organs for transplant. Some states have in addition passed laws requiring medical personnel to ask the family of the deceased to donate his or her organs. Thus, we have given social approval to a system in which needed organs are donated but not to systems in which they are routinely taken without permission or sold as commodities on the open market. Nevertheless, as I noted at the outset of the chapter, this system of giving and receiving has not provided as many donated organs as are desired for transplant purposes.

On some occasions organs are given by *living* donors, but this can be permitted only within clear limits. Years ago Paul Ramsey called attention to one of those limits, recounting the following fictitious case study:

> Many months ago the 15-year-old son of Mr. Roger Johnson was admitted to a Houston, Texas, hospital for tests to determine the cause of his generally debilitated condition. Use of the latest available diagnostic techniques and equipment eventually led to the conclusion that the lad was suffering from a progressively deteriorating congenital condition of the valves of the heart. The prognosis communicated to the distraught Mr. Johnson was that his son could not live past the age of 20, and that there was no known treatment for the malady with which he was afflicted.
>
> At first Mr. Johnson tried to resign himself to his son's plight. Then he began to brood and think of the pleasures and joys of adult life which he, at the age of 42, had already known, but which his son would never know. The more he thought of this, the less willing he became passively to accept the doctors' verdict. Finally he thought of a means by which his son's life might be spared.
>
> His plan, which he communicated to a physician friend, was an uncomplicated one. In light of the success of recent heart transplant operations with unrelated donors and donees, he reasoned, there must be a high probability that a transplant of the heart of a genetic relative would be successful. Accordingly, he would simply donate his own heart to his son. He had lived a full life, he said, and he could leave his son well provided for financially. His wife had died several years earlier, so that complication was not present. His own parents had no rightful claim to his continued life. He asked his friend's aid in finding a physician who would perform the operation. Not without considerable misgivings, his friend complied,

> eventually finding a heart surgeon eager to attempt the transplant of a heart from a healthy and related donor not *in extremis* at the time of the operation.
>
> In the course of preparation for the transplant, elaborate precaution was taken to ensure that the son would not know the real nature of the proposed operation. He was told simply that a transplant operation on his heart was to be attempted in the hope of prolonging his life, and he agreed to try it with full knowledge that death could certainly result if the try were unsuccessful. In reality, of course, it was contemplated that Mr. Johnson's heart would be removed from his chest while he was under general anaesthesia and that it would be transplanted in the chest cavity of his son.
>
> When the date of the scheduled operation arrived, the father went to the son's room, affectionately wished him good luck, and returned to his own room to be prepared for his own operation. He was eventually placed under general anaesthesia, and taken to a special operating room to await the transfer of his heart to an oxygenating and circulating "heart-lung" machine.
>
> He is in the operating room now, and the surgeon is scrubbing. You are chief of staff in the hospital in which the operation is to take place. You had no prior knowledge of the operation, but this is frequently so. A worried nurse has brought you word of the planned operation on this occasion. You have power to stop the operation. Should you do it?[3]

The case is striking because it makes clear what Christian rhetoric about "love" and "freedom" sometimes blurs: Not every gift can properly be given by those who know themselves to be creatures rather than Creator. The body, as the place of personal presence, has its own integrity, which ought to be respected. Indeed, because we are re-

3. Paul Ramsey, *The Patient as Person* (New Haven and London: Yale University Press, 1970), pp. 18-90.

garded as stewards rather than owners of our bodily life, the Roman Catholic and Jewish traditions generally forbade self-mutilation. These traditions have become willing to approve the self-giving of organs or tissues for transplantation as long as the donation will not cause grave harm to the donor's bodily life. Certainly any solid organ donation — such as that of heart, liver, or lung — that would cause death or great harm to a living donor is not a proper work of *creaturely* love. (Interestingly, an increasingly secular society, in which many people do not share Christian and Jewish disapproval of suicide, may find it hard to explain why such donations should be forbidden or why the case study recounted from Ramsey should remain fictitious.)

In general, therefore, we may regard donation of a kidney or of bone marrow as significantly different from donation of heart, lung, or liver. (In recent years partial grafts of liver and lung tissue, which do not involve transplantation of the entire organ, have been done. Presumably we should think of them as more like bone marrow donation.) Yet, a living donor's gift even of tissue or a paired organ (such as the kidney) should not simply be affirmed as if it were morally uncomplicated. Doctors have in the past been hesitant to transplant kidneys from living, *unrelated* donors, and it is good that they should be. We should want them to be reluctant to subject a healthy person to the risks of a major operation and the loss of one kidney even if that person is eager to make this bodily gift. It is true, of course, that we ought always be ready to risk harm to ourselves for the sake of others. But it is one thing to aim at my neighbor's good, knowing that in so doing I may be harmed; it is another to aim at my own harm *in order* to do good to my neighbor. Thus, even when we approve donation (of, for example, a kidney) from a living donor, we should retain a lively sense of the moral complexity of such an act.

In recent years the number of kidney donations from living, *unrelated* donors has increased.[4] In part this has been due to a growing will-

4. Gina Kolata, "Unrelated Kidney Donors Win Growing Acceptance by Hospitals," *The New York Times,* 30 June 1993, p. B6.

ingness to accept donation from the spouse of a patient suffering from kidney disease, but there have also been cases of such donations between friends or even, simply, acquaintances. In the last decade there was a tenfold increase in the number of biologically unrelated living donors.[5] Though obviously beneficial in certain respects, this may also be troubling if — in depicting generosity in ways unmoored from the body and the body's connections — it suggests a separation of who we are from the body that we are. Because the increased willingness to permit such donations is due in part to the pressure for organs and the desire of transplant surgeons to do what they can to meet that need, we must beware of the tyranny of the possible, the pressure to suppose that we ought to do whatever we are able to do. Bioethicists have generally worried that unrelated donors might be pressured or paid, or that spouses might feel a kind of pressure that keeps their consent from being truly free. No doubt such concerns are legitimate and deserve our attention. But consent is not the only moral issue here, and those concerns should not obscure a larger underlying issue: the integrity of bodily life. If we learn to regard our bodies simply as collections of organs potentially useful to others (and available whenever our true inner self chooses to give them), we are in danger of losing any close connection between the person and the body. That connection has always been affirmed in Christian thought, although it has often been a fragile connection. We are regularly tempted to suppose that the "real" person transcends the body. When we do that, dehumanization lies near at hand. An acute sense of that dehumanizing tendency to regard our bodies as collections of alienable parts moved Leon Kass to refer to organ transplantation as "a noble form of cannibalism."[6] That striking phrase is not overdone as long we take the whole of it seriously. Not just cannibalism, but *noble* cannibalism. Kass would not have us ignore

5. 2002 Annual Report of the U.S. Organ Procurement and Transplantation Network.

6. Leon R. Kass, "Organs for Sale? Propriety, Property, and the Price of Progress," *The Public Interest*, Spring 1992, p. 73.

the nobility involved in gifts of the body, but neither would he have us think too casually about the body's own integrity and its meaning as the place of personal presence.

Because of long-standing reservations about organs given by living donors and notwithstanding the recent rise in donations from biologically unrelated living donors, the tendency in transplantation (since the discovery of drugs to suppress the body's immune reaction that rejects foreign tissue) had been to use cadaver organs taken immediately after death. (This assumes, of course, that the deceased had, while still living, authorized such donation, or that appropriate family members had done so after his death.) And, of course, from a cadaver one can take for transplant not only a paired organ such as the kidney but unpaired organs such as the heart. Why not encourage Christians to make such gifts of the body?

Yet, even here a certain caution is in order. Given the increasing pressure to make more organs available for transplant, we will see a growing tendency to think of cadaver organs as a communal resource available for the taking — unless perhaps the family of the deceased objects. That tendency ignores the human significance of burial — of a family's desire to take leave of a loved one. William F. May has noted that it is "wrong, indecorous, and enraging" to force a family "to *claim the body as its possession,* only in order to proceed with rites in the course of which it must acknowledge the process of surrender and separation."[7] May recalls a tale from the Brothers Grimm in which a young man, who is incapable of horror and does not shrink back from the dead, attempts even to play with a corpse and is sent away "to learn how to shudder." If, as I noted at the outset of this chapter, families are often reluctant to authorize organ donation after death of a loved one, that reluctance ought to be honored — lest we collectively forget how to shudder. Indeed, I do not think it wise even to act upon the deceased

7. William May, "Attitudes toward the Newly Dead," in *Death Inside Out,* ed. Peter Steinfels and Robert M. Veatch (New York: Harper & Row, 1974), p. 144.

person's previously stated willingness to be a donor in the face of family reluctance or objection. His very corpse from which the organs would be taken is the best evidence that there are limits to his freedom to determine the course of his life or the disposal of his remains.[8] Our society's desperate attempt to find ways to live longer should not be allowed to override a deep-seated and difficult to articulate sense of the importance of the body, even the dead body.

Slippery Slopes

When cyclosporine, the first powerful immunosuppressive drug, was discovered in 1972, transplantation technology was revolutionized. If the immune system's rejection of an alien organ could be overcome, the possibilities seemed endless. No longer would transplants be conceivable only if donor and recipient were closely enough related to be a good match. And once donation from strangers became reasonable to contemplate, it also became possible to move beyond living donors' gifts of paired vital organs (such as a kidney) to transplantation of unpaired vital organs (such as the heart or liver) from cadaver donors. The crucial conceptual notion here was that of "brain death."

In 1968 an *ad hoc* committee at Harvard recommended a neurological criterion — cessation of brain activity — for determining death. Prior to that, cessation of heart and lung activity — a cardiopulmonary criterion — had been generally used to mark the point of death. But it had by then become possible to sustain heart and lung activity (with a respirator) for days or even weeks after a patient had irreversibly lost all brain function. Therefore, the two traditional "vital signs" of heart and lung activity could be maintained solely through mechanical assistance. In these circumstances it made sense

8. Leon Kass, "Organs for Sale?" p. 74.

to many to say that a human being actually dies when brain activity ends, because only that activity makes possible the body's ability to function as an integrated whole. The Harvard committee attempted simply to fix criteria on the basis of which physicians could determine that a patient was neurologically dead. Its criteria — including lack of responsiveness, no breathing or movement (when off the respirator), no reflexes, and a flat EEG — have been largely accepted in the years since then. This position received powerful support in a 1981 report — *Defining Death* — released by the President's Commission for the Study of Ethical Problems in Medicine and Biomedical and Behavioral Research, and criteria like those above have been written into law in our states. These criteria were intended to determine when *all* brain activity had ended, when "whole brain" death had occurred. A person can, of course, suffer the loss of "higher" brain (cortical) function, losing the capacities for awareness or self-consciousness, while brain stem functions (controlling spontaneous breathing, eye-opening, etc.) remain. According to these criteria, loss of higher brain functions alone would not constitute death, and the laws of our states that have established criteria for determining brain death have had *whole* brain death in view.

Among the ironies of our current situation is this: Just when the "whole brain" understanding of death seemed to have become increasingly established both in law and our common morality, it has come under powerful critical scrutiny from scholars who think it mistaken (and, indeed, who may suspect that it was less a theoretical advance than a practical accommodation to the needs of improving transplantation technology). The irreversible loss of all brain activity had been thought to be an adequate criterion for marking somatic (bodily) death because the brain serves to integrate and coordinate the body's activities. But, in fact, there *are* integrative functions of the body that are not simply coordinated by the brain. For example, homeostatic regulation of bodily fluids, control of body temperature, and infection fighting can be found at least for a time in some patients after they have experienced "whole brain" death.[9] This is an argument that has yet to run its course. That some bodily functions can be artificially replaced or continue for a time in brain-dead patients does not necessarily disprove the claim that it is the brain that coordinates the body's activities and thus sustains (or fails to sustain) the self-directed coordination of parts for the sake of the whole organism. Until we achieve greater clarity, we are likely to continue to work with the criteria for determining whole brain death.

This is not unreasonable, but we have to grant that the concept of "brain death" remains conceptually and experientially puzzling in some ways. It permits transplant surgeons to retrieve the organs of a neurologically dead person while, because of mechanical assistance, circulation of oxygenated blood sustains the vitality of those organs in the "corpse." Yet, of course, even if we agreed that irreversible loss of whole brain function established that the person was dead, we would be reluctant to bury a corpse until its heart had ceased to beat. We seem willing, therefore, to remove organs for transplant from a corpse before we would be willing to bury it. The body has died, because it can no longer function as an integrated whole; yet, with mechanical assistance some organs and tissues, taken by themselves, retain vitality. If that makes us uneasy, we might prefer to remove mechanical assistance and let the body die "all the way." But then, of course, its organs are unlikely to be usable for transplantation.

More than a quarter century ago, when this move to "update" criteria for determining death began, it was met with suspicion. At that time the technology of transplant surgery was beginning to make progress, and some people suspected that the desire to establish in law a concept of brain death was motivated only by the wish to obtain organs for transplant before those organs had deteriorated (as they will

9. D. Alan Shewmon, "The Brain and Somatic Integration: Insights into the Standard Biological Rationale for Equating 'Brain Death' with Death," *Journal of Medicine and Philosophy* 26 (October 2001): 457-78.

rapidly when heart and lung activity fail). In truth, however, there were other reasons — apart from the desire for transplantable organs — to rethink the criteria for determining death, since one needed to decide whether a respirator was simply oxygenating a corpse or sustaining a living human being.

The suspicions may not have been entirely groundless, however — or, perhaps better, they may have been ahead of their time. For it has become clear in recent years that the thirst for transplantable organs is so strong that we *are,* in fact, tempted to redefine death *in order* to secure the "needed" organs. For example, in 1994 the Council on Ethical and Judicial Affairs of the American Medical Association issued an opinion holding that it is "ethically permissible" to use "the anencephalic neonate" as an organ donor, even though, as the Council recognized, under current law anencephalic babies are not dead.[10] Anencephaly is a condition in which an infant is born with a fully or partially functioning brain stem but without any cerebral hemispheres (higher brain). These infants can never have any awareness of their own existence or of the surroundings in which they live, and they usually die within hours or days. With aggressive treatment it may on occasion be possible to sustain their life somewhat longer, but, because they are essentially dying patients, it seems better simply to give them what care and comfort we can while permitting them to die without the bodily intrusiveness of aggressive measures.

It is worth noting that as recently as 1988 the AMA's Council on Ethical and Judicial Affairs had concluded that it was not permissible to remove organs for transplantation from anencephalic infants while they were still alive, even though it is harder to maintain organs in suitable condition if one waits until the infant has sustained whole brain

10. Council on Ethical and Judicial Affairs, American Medical Association, "The Use of Anencephalic Neonates as Organ Donors," *Journal of the American Medical Association* 273 (May 24/31, 1995): 1614-18. It is worth noting the anesthetizing quality of the Council's language. "Anencephalic neonates" are, in fact, severely disabled newborn babies — perhaps the weakest members of the human community.

death. The Council's more recent opinion is quite frankly based on a sense that it is imperative to acquire organs for transplant.

> Newborns and other young children usually can benefit from organ transplants only if the organs are taken from children of similar size. However, there is a serious shortage of pediatric organ donors. As a result, each year approximately 500 children need heart transplants, another 500 need liver replacements, and approximately 400 to 500 children in the United States need kidney transplants. With the scarcity of hearts, livers, and kidneys available for transplantation, 30% to 50% of children on the transplant waiting list die while waiting for a suitable organ. These figures are undoubtedly underestimates of the shortage of pediatric organs. With the long waiting lists for the organs, many children in need never make it onto the lists because they would not have high enough priority to receive an organ or because they do not live long enough to have their names entered on the waiting list.

For these reasons the Council approved what we would ordinarily regard as wrong. Normally, an unpaired vital organ such as the heart could be taken for transplant only from a *cadaver* donor (who had previously consented or whose family had consented). But within only six years the Council reversed its earlier position and approved such "donations" from anencephalic infants — approved, we should not hesitate to say, taking the life of these infants in order to make their organs available for transplant to other children whose life prospects are better. "Permitting such organ donation," the Council suggests, "would allow some good to come from a truly tragic situation, sustaining the lives of other children and providing psychological relief for those parents who wish to give meaning to the short life of the anencephalic neonate."

It happens that in December 1995, the AMA's Council, under considerable pressure from its House of Delegates, once more reversed direction and rescinded its 1994 opinion permitting organ donation from living anencephalic infants. It did so, however, only on the ground that doubt had arisen whether all anencephalic infants lack consciousness and whether an assured diagnosis of anencephaly is always possible — that is, on purely technical grounds. Therefore, if further study were to demonstrate that these infants do lack consciousness and that their condition can be reliably diagnosed, the Council would have no reason not to change direction once again and approve the use of living anencephalic infants as organ donors.

This is the sort of slippery slope on which we stand if we permit ourselves to believe that ours is the godlike responsibility of bringing good out of every human tragedy. We suppose that the task of giving "meaning" to a child's life is ours, and we permit ourselves to use the infant's death as a means of psychological relief for others. Moreover, we will gradually learn to think of ourselves and others not as living beings whose bodies have their own unity and integrity but as "ensembles of parts . . . to be given away or taken or — worst of all — sold."[11] We are on the way to seeing ourselves, in Paul Ramsey's arresting phrase, as "a useful precadaver." That I do not exaggerate can be seen from procedures for procuring organs for transplant from what are called "non-heart-beating cadavers."[12]

As I noted above, most organs for transplant come today from cadaver donors who have been declared neurologically (brain) dead but whose hearts are still beating because of mechanical assistance, thereby sustaining the vitality of their organs before transplant. Because the supply of donor organs does not meet demand, however, the search is always on for new sources of organs. In recent years, the University of Pittsburgh Medical Center, a major center of transplant surgery, was the first to focus that search on non-heart-beating cadaver donors. These are patients who have been declared dead by traditional

11. Paul Ramsey, *The Patient as Person*, p. 208.

12. The *Kennedy Institute of Ethics Journal* 3 (June 1993) contains articles from a variety of perspectives examining this issue.

cardiopulmonary criteria after they or their families have decided to forgo any further treatments. After the decision to forgo further life-sustaining treatment has been made, the still living person is taken to the operating room. There therapy is withdrawn, the patient dies on the operating table, and his organs are removed immediately after death is declared. Objecting to this on a variety of grounds, Renée Fox, a sociologist whose pioneering studies of transplant technology are well known, has singled out as "most dreadful" what she terms "the desolate, profanely 'high tech' death that the patient/donor dies, beneath operating room lights, amidst masked, gowned, and gloved strangers, who have prepared his (her) body for the eviscerating surgery that will follow."[13] Perhaps if our noble desire to prolong life leads us to such ignoble means, we need to be sent away to learn how to shudder.

Rather than shuddering, it is of course possible to forge boldly ahead. If the Pittsburgh Protocol for obtaining organs seems almost to mock the view that unpaired vital organs should be taken only after the donor has died — to mock it, that is, by adhering to the letter but not the spirit — we might instead simply abandon the claim that it is always necessary to wait for death before procuring organs for transplant. Without recommending it, Robert Arnold and Stuart Youngner describe what this might mean.

> Machine-dependent patients could give consent for organ removal before they are dead. For example, a ventilator-dependent ALS patient could request that life support be removed at 5:00 p.m., but that at 9:00 a.m. the same day he be taken to the operating room, put under general anesthesia, and his kidneys, liver, and pancreas removed. Bleeding vessels would be tied off or cauterized. The pa-

13. Renée C. Fox, "'An Ignoble Form of Cannibalism': Reflections on the Pittsburgh Protocol for Procuring Organs from Non-Heart-Beating Cadavers," *Kennedy Institute of Ethics Journal* 3 (June 1993): 236.

> tient's heart would not be removed and would continue to beat throughout the surgery, perfusing the other organs with warm, oxygen- and nutrient-rich blood until they were removed. The heart would stop, and the patient would be pronounced dead only after the ventilator was removed at 5:00 p.m., according to plan, and long before the patient could die from renal, hepatic, or pancreatic failure.
>
> If active euthanasia — e.g., lethal injection — and physician-assisted suicide are legally sanctioned, even more patients could couple organ donation with their planned deaths; we would not have to depend only upon persons attached to life support. This practice would yield not only more donors, but more types of organs as well, since the heart could now be removed from dying, not just dead, patients.[14]

Arnold and Youngner do not, as I noted, claim that we should turn in this direction, but they view it as an honest projection of where we may gradually be headed.

In recent years we have also seen stories of children conceived *in order* to serve as bone marrow donors for family members. Increasingly, some argue that we should permit the sale and purchase of organs needed for transplant — that, in this way at least, the body may be a commodity for sale. Having started down the highway of transplantation, we seem unable to find any exit ramp as we press toward a vision of humanity in which everyone becomes "a useful precadaver."

Can our public policy find that exit ramp? Not unless we first recover it for ourselves. The truth is, we will do almost anything to keep ourselves or our loved ones alive. Whatever we may think public policy ought to be, if our own life or our child's were at stake, we might well bend all our energies to the task of finding an organ for transplant.

14. Robert M. Arnold and Stuart J. Youngner, "The Dead Donor Rule: Should We Stretch It, Bend It, or Abandon It?" *Kennedy Institute of Ethics Journal* 3 (June 1993): 271.

Whatever could be done we would be tempted to do, and we are therefore helpless in the face of the relentless advance of this technology. Christians, who know that death is indeed an evil and the last enemy opposed to God's will for the creation, should find the temptation quite understandable. But we also need to develop the trust and the courage that will enable us sometimes to decline to do what medical technology makes possible. Only by supporting organ transplantation in ways that do not lose the meaning of the body as the place of our personal presence, and in ways that do not imply that staying alive as long as possible always has moral trump, can we become people who give thanks for medical progress without worshiping it or placing their trust in it. In becoming such people we may bear a different kind of life-giving witness to our world.

CHAPTER TEN

Gifts of the Body: Human Experimentation

In her "Personal Health" column in the *New York Times,* Jane Brody once commented on the difficulties researchers have in recruiting people to serve as subjects in clinical experiments.[1] Despite the impressive gains made by modern medical research and the promise of still greater advance on the horizon, "only a small percentage of those eligible participate in such trials." The clinical trials to which she refers are generally "randomized." That is, some of the subjects participating will receive the experimental therapy being studied, while other participants (who serve as a control group) may receive only a placebo having no known pharmacological effect. Participants do not know which "treatment" they are, in fact, receiving, nor are they given any choice in the matter. Neither, however, do the researchers themselves make the choice; instead, it is randomized. Without the use of such a control group, the results of research can seldom be considered reliable, for researchers must have some confidence that the effects they see are actually due to the experimental therapy rather than to other factors.

We can, of course, understand why some patients might be reluctant to participate in such trials, even if they face life-threatening ill-

1. Jane E. Brody, "Personal Health," *The New York Times,* 16 November 1994, p. B8.

ness for which standard therapies have been ineffective. Even in dire straits, we might be reluctant to become less a "patient" than a "research subject," losing control over the treatment we receive. So it is perhaps not surprising that Jane Brody should report that "only a small percentage of those eligible participate in such trials." That low rate of participation may be disappointing, but it is not surprising. Much more striking, however, is another fact that she reported. "Interestingly, while children with cancer account for less than 2 percent of all cancer cases, about 60 percent of them participate in clinical trials. (Treatment of childhood cancers is complex and often not readily handled by community doctors.) But for adults with cancer, an overwhelming majority of cancer patients, only about 2 to 3 percent enroll in clinical trials."

That statistic ought to catch our eye. Those old enough to decide for themselves whether to participate are quite reluctant to do so. But we seem far less reluctant to enroll children who cannot decide for themselves. Perhaps the difference is explained in part by the fact to which Brody points. Perhaps to get any treatment at all for childhood cancers one must often go to a major research center where treatment is available only within a clinical trial. Perhaps the desperation of parents seeking any means of saving their child's life also accounts for the high percentage of children who are enrolled as research subjects. One need not be a cynic, however, to wonder whether we seem a little too willing to enroll children who cannot speak for themselves in undertakings we ourselves avoid. Such a possibility is worth at least contemplating as we reflect here upon the ethics of experimentation.

Therapy and Research

Beyond any doubt we have reaped enormous benefits from modern medicine, based, as it is, upon scientific experimentation. We need only recall a disease such as polio, so feared only forty years ago and now so rare in this country, to remind ourselves what medical research has done for our lives. Aimed not primarily at the care of particular patients but at the acquisition of generalizable knowledge that may help future sufferers, research has radically altered the landscape of medicine. It has also, however, transformed our understanding of medicine's place in life. In the Hippocratic Oath a physician swore to "apply dietetic measures for the benefit of the sick," and that *patient*-centered medicine remains of great value to us. Its desirability helps to explain why relatively few adults enroll as experimental subjects. They want to be *patients*, with *physicians* focusing on their benefit, not *subjects* used by *researchers* in the cause of medical progress. At any time, but perhaps especially when we are ill, we do not want to think of our doctor simply as a public functionary serving some larger social good.

But we also are not eager to give up the medical benefits research has brought us. Of course, even traditional Hippocratic medicine that is focused on a patient's well-being will often have an experimental aspect. Physicians might try standard therapies without any success and be moved to suggest, as therapy of last resort, treatment that could only be termed experimental. Moreover, physicians almost always gain some knowledge simply by trial and error. For the experienced physician as much as for the intern, all medical practice does indeed involve "practice." Scientific medicine, however, seeks to advance beyond knowledge gained in trial-and-error fashion and aims to give both physicians and patients confidence that what is offered as therapy genuinely has the weight of evidence behind it.

Just as therapy sometimes — or even often — has an experimental aspect, so too research subjects may sometimes or often hope that theirs will be the fortunate case in which an experiment not only produces knowledge that will bring benefits in the future but also helps them. Because therapy and experiment are more easily separated in thought than in practice, and because consent of the patient/subject is needed in either case, one might argue that it is unwise to make any sharp distinction between them.[2] I think, however, that the difference between therapy and experiment continues to be of moral importance. After all, in every instance I as a patient may well want to ask my doctor: "For whose sake primarily am I undertaking this regimen? My own? Or that of future sufferers?" If there is at present little evidence to indicate that the regimen is likely to help me, I may, of course, still hope that it will. But lacking such evidence, I should not hide from myself the truth that I am acting chiefly as an experimental subject — assisting in the search for generalizable knowledge that may bring benefit to others in the future. Facing that truth squarely, I will be better able to evaluate the meaning and worth of the undertaking.

Any serious Christian evaluation of the place of medical research must reckon with the human tendency toward idolatry. We can — and probably have — made an idol of medical advance. Noting the tension between scientific medicine (which wars against death) and clinical medicine (which knows that every patient dies), Daniel Callahan has mused upon a paradox in our attitude toward death. Although we grant that death is inevitable, we do not admit that any medical cause of death cannot be overcome. The paradox is likely to incline us feverishly in the direction of experimentation and research.

> The goal of scientific-research medicine is to overcome disease, to overcome that which brings illness and death. Death is the greatest enemy, and disease its army. There is no important human disease, or class of diseases, that the National Institutes of Health are not seeking to cure and eliminate. . . . Every physician knows that patients die, that for every illness for which a cure can be found for a patient at this time, there will be one, at a later time, that will not admit of a cure. . . . For all the obviousness of that perception, however, scientific medicine has managed to keep it at bay in its own

2. Nancy M. P. King, "Experimental Treatment: Oxymoron or Aspiration?" *Hastings Center Report* 25 (July-August 1995): 6-15.

> research logic. . . . No cause of death has been declared beyond hope; none could be. All of the known causes of death can, in principle, be picked off, one by one.[3]

Pondering Callahan's claim, Christians ought to feel ambivalent. On the one hand, we know that death is truly an enemy — an enemy so powerful and with so legitimate a claim upon us that God himself had to taste it in order to redeem us from it. That Christian perception of death as enemy has surely been one of the motive forces in our civilization's drive to overcome death — a drive that is represented in the goals of research medicine. We ought not, therefore, be too ready to accept death simply as a natural part of life. We ought not too easily accept the death of any human being; for each person is made for God, unique and irreplaceable in that relation. Death is always an enemy that threatens to separate us from the One for whom we are made.

If we ought not acquiesce too readily in the direction of Callahan's argument, we should however grant its force. His reading of our contemporary situation demonstrates clearly our idolatrous propensities. Fearing death and running from it, we find no good reason ever to accept it in our own life or that of others. Placing our hope in the forward march of medical research, we deceive ourselves into imagining that it could be redemptive, that it might overcome the sting of death. In short, we fashion the golden calf of research medicine.

Our evaluation of the place of medical research in human life ought to reflect our understanding of the human person as both finite and free. As free spirits made for God, we know death as an enemy which threatens to cut us off from the One for whom we long. It is an enemy against which we rightly set our face. As finite creatures made from the dust of the ground, we also know that our longing for God is not simply a longing for more of this life — and that at some point, for

3. Daniel Callahan, *The Troubled Dream of Life: Living with Mortality* (New York: Simon & Schuster, 1993), pp. 73-75.

each of us, death must be acknowledged. Our evaluation of the place of medical research in human life ought also to reflect our understanding of the meaning of suffering. It is a great evil from which Jesus himself shrank back, and we ought to do what we can — including engaging in medical research — to relieve it. But the march of progress within human history is not itself redemptive, and God ultimately deals with suffering in his own mysterious way. God bears it into death — demonstrating thereby that no other gods of our own making, however powerful they may seem, can deal sufficiently with the suffering that marks our lives. Christians therefore have no good reason to renounce the cause of medical research, but our commitment to it ought to be a chastened one, liberated from the fear that makes an idol of our hopes. Whether our society can manage this is difficult to know, but we ought to help shape such a chastened understanding of research medicine.

Participation in Research

The terrible story of medical experimentation carried out by the Nazi regime in the mid-twentieth century has been instrumental in shaping our research ethic. At the heart of that ethic is the requirement of *consent*. Research subjects must be able to consent and should not be enlisted or used without their consent.

Why should anyone give such consent? Why should anyone be encouraged to serve as a research subject? As I noted in the opening chapter, Christian vision depicts our nature as both finite and free. As limited, finite beings we quite naturally look to our own interests and those of others closely bound to us by ties of kinship and love. Because, however, we are also made for God, we always transcend those limited attachments. We are able to place ourselves in the position of others who are less closely bound to us, able to consider and serve their needs also. If I participate in research that is unlikely to benefit me but that may help future sufferers, I need not deceive myself with the vain hope that mine may be the exceptional case in which benefits come immediately. Rather, this is one way of serving the "stranger" — that unknown future sufferer to whom I am not naturally or immediately drawn. If Christians are to love any neighbor who is in need, this may be reason enough to consider participating in research when an appropriate occasion arises.

Research is not an individual undertaking, however. It is a social project, carried out with society's resources. For that reason we should do more than encourage ourselves and others to consent to participate. Consent must also be a requirement for participation. Knowing how quick we are to use some people for the good of others, how readily we may conscript unwilling subjects into a cause they have not made their own, we can guard against that tendency only by assuring ourselves that research subjects have given their consent — and that they have done so freely and knowledgeably. No one can truly be said to be a volunteer in the cause of medical progress if, for example, he is led to hope that he may benefit from research when such a result is highly unlikely. We cannot expect our society's ongoing research endeavor to be a blessing for us if it is built on false hopes or on participation by subjects whose consent is doubtful.

If we take the consent requirement seriously, we will have to think again about the statistics noted at the outset of this chapter. Only a small percentage of adults eligible to participate in randomized clinical trials actually do so; yet, "while children with cancer account for less than 2 percent of all cancer cases, about 60 percent of them participate in clinical trials." Those who cannot speak for themselves, those who cannot freely and knowingly give their own consent, "participate" in much larger percentages than the rest of us. Do we build the cause of medical progress on their backs? Our society has gradually drawn back from a strict interpretation of the Nuremberg Code that was formulated after World War II. That code, were its requirement of consent taken strictly, would have made it impossible to use children as research subjects. We have gradually come to accept the proxy "consent" of parents as sufficient warrant for using children in research. Perhaps under some circumstances it is, in fact, sufficient. For example, a good bit of research involves no discernible risk to those who participate in it. Enrolling their child as a participant in that sort of research may be one way by which parents begin helping a child understand what it means to care about the good of others. Proxy consent in such circumstances seems reasonable. But we are all too ready to use those who cannot speak for themselves, and if the percentage of children participating in research is considerably greater than that of adults, we ought to stop and ask whether the language of proxy consent has not become deceptive and illusory. Moreover, children suffering from serious illnesses should not be burdened in their dying by participation in research — as if, contrary to the fact, they could "volunteer" to be soldiers in the cause of medical progress. The knowledge that might be gained from their participation is knowledge we should do without or acquire in slower, trial-and-error fashion.

Nor are children the only weak and vulnerable subjects whom we might be tempted to exploit. If we take the consent requirement seriously, we will need to worry, for example, about a report (from 1996), detailing how a homeless man in Indianapolis earned $2,400 for spending eight weeks as a subject in a Phase I drug trial conducted by Eli Lilly & Co.[4] The report notes that since 1988, 94 percent of Lilly's volunteers have listed permanent addresses on their applications — though this information is not necessarily verified. So it may be difficult to judge for sure how many such homeless people are bearing the burden of testing the drugs others of us will one day use. And, of course, it is possible to argue that, rather than experiencing it as a burden, they see it as desirable, understand what they are doing, and should not be deprived of the opportunity. Still, that is an interpretation the rest of us are probably all too ready to accept, and we ought to be at least a bit suspicious of our eagerness.

4. Rick Callahan, "Drug Company Defends Testing on the Homeless," *The Times* (Northwest Indiana), 21 November 1996, p. E1.

In short, in the second half of the twentieth century we have, under the pressure and the lure of scientific advance, moved away from an older understanding of medicine focused on the good of patients alone (for which all research was simply retrospective trial-and-error learning). Scientific medicine has made it possible for us to put nature to the test, and we have gradually come to believe that such testing is imperative — that because we *can* gain knowledge that will help future sufferers, we *must* do so. The requirement of consent was developed both to make research possible and to limit it. To make it possible — by authorizing us to enlist those who truly volunteer. To limit it — by guarding against our tendency to use some, without their full consent, for the good of others. Because the consent requirement sets limits to the advance of research, we will always be tempted to find ways around it.

The only protection against that temptation — which lures us only because we are at heart idolaters — is the certainty that *how* we live is more important than *how long,* that what we *do* is finally of more weight than what we *accomplish,* that there are limits to our responsibility to relieve suffering. Protected by such an understanding, we can be grateful for the benefits of research medicine and hope for continued progress without making of it a god. In the ancient myth of Prometheus, all the other animals had been given forms of protection — swiftness, fur, wings, shells. No protective covering was left for human beings, until Prometheus went to the sun, lit a torch, and brought fire to the earth. He gave to human beings a special gift — a kind of knowledge that made technological advance possible. We can welcome that knowledge with grateful hearts as long as we also remember its ambiguous character. As Prometheus gives the fire, a satyr stands near wanting to embrace it, and Prometheus has to warn him: "It burns when one touches it, but it gives light and warmth, serving all crafts providing one knows how to use it well." That ambiguous lesson is well worth remembering as we contemplate both the urgency and the limits of our responsibility for continued medical progress.

CHAPTER ELEVEN

Embryos: The Smallest of Research Subjects

One special problem for research ethics deserves separate attention — in part because it has been the subject of great controversy in the United States (and elsewhere in the world) in recent years, and in part because it forces Christians to think about questions that bring their deepest beliefs to the surface. This is the problem of research on embryos, and, in particular, research using embryonic stem cells derived from the inner cell mass of early embryos, which are destroyed in the process.

The reasons why we might be eager to conduct such research are neither hard to find nor hard to understand. In addition to the undeniably powerful desire to expand our knowledge of cell differentiation and organismal development, surely a powerful motive in itself, there is hope that this research might open striking new avenues of treatment in regenerative medicine — bringing cures, or, at least, help through transplantable cells for patients suffering from such debilitating conditions as Parkinson's disease, spinal cord injury, or diabetes.

Stem cells hold such promise because they are able to renew themselves for a prolonged period of time and because they are able to produce more differentiated cell types that make up the tissues and organs of the body. There are stem cells in the body that are not *embryonic* stem cells; indeed, bone marrow stem cells have been used for years in treating cancer. Derived from blood, brain tissue, muscle, and many other sources in the body, these are sometimes called "adult stem cells," though the term is misleading in some respects and designates simply *non*-embryonic stem cells. What makes the case of embryonic stem cells special is that, on the one hand, they are thought by many to have greater potential for differentiation into a wide range of tissues, and, on the other hand, procuring them requires destruction of the embryo.

Because adult stem cells may hold out significant treatment possibilities, saying no to embryonic stem cell research does not mean saying no to all stem cell research with its possibilities for regenerative medicine. Nevertheless, the relative usefulness of adult and embryonic stem cells is disputed, and it may well turn out that the latter have more potential for cures than the former. That alone would not settle the matter, however. If intentional destruction of embryos in the course of research is wrong, then we would simply have to accept the fact that progress toward cures would have to come more slowly and haltingly. We are not, after all, obligated to use any and every means to pursue the curative goals of medicine. All honor, of course, to the humanitarian motives that are present in research, and we should not underplay the moral mission of medicine to heal, from which, indeed, all of us have profited. But that does not mean that this mission should proceed without any limits.

It is also worth noting at the outset that the embryos from which stem cells are derived may themselves be produced in different ways. Some embryos may result from fertilization in the laboratory — whether produced specifically for research or produced (but then not needed) for couples seeking treatment for infertility. It may also be possible to clone embryos specifically for use in research; indeed, this is what researchers would prefer to do right now, since it would permit them to create the precise disease models they want to study. These several different sources of embryos complicate the discussion considerably. The possibility that cloned embryos could be used means that two concerns are intertwined — the destruction of embryos for research, and crossing the boundary that divides sexual from asexual reproduction in the human species. The possibility that embryos could be produced specifically for research or produced for infertility treatments (and then become available for research when they are no longer needed for reproductive purposes) means that we need to ask whether this difference in original purpose makes any difference in the moral judgments we should make.

Destroying Embryos

Each of us began life as an embryo, for an embryo is a human being in its earliest stage of development — fragile and undeveloped, to be sure, but nevertheless an integrated, self-developing whole. As such an integrated organism an embryo is capable (though by no means assured) of the continued development that characterizes human life. There is certainly something mysterious about an embryo. It hardly resembles the human beings with whom we spend our days (although, of course, it looks very much as they looked when they, too, were in the earliest stages of their lives). It lacks most of the characteristically human traits that we treasure (though it has them in potential and, in that sense, has them even now).

Focused as we tend to be on these characteristics and capacities, we forget that there are different points in the trajectory of any human life. There is a beginning, constituted largely by potential as yet undeveloped. There is what we may call a zenith, when the organism is at the height of its powers. And there is a decline, during which only some residue of those distinctive capacities remains. But none of these moments *is* the human being. On the contrary, a human being is a single organism with a continuous history. Beginning in that early embryo and continuing — alas and perhaps — through irreversible coma, every human life is a single personal history.

Advocates of embryo research have always struggled with these facts. They have not wanted to say that embryos are simply available for use by the rest of us, and so they have generally said that embryos should be treated with "respect" — or, even, "profound respect." Although I do not doubt that it is under certain circumstances possible to show moral respect for a human being whom one kills, it is not easy to take seriously the language of respect when we bring an embryo into existence for the sole purpose of using (and destroying) it in research. (And this would, it is worth noting, always be the case with cloned embryos. Since almost everyone, at least for now, opposes implanting, gestating, and bringing to live birth cloned embryos, the only point of cloning embryos would be to use them in research.) In these circumstances it would be far more honest simply to drop the language of respect entirely.

Moreover, we need to remind ourselves what the phenomenon of embryo research would actually involve. We are not talking about whether it might be possible — even in an anguished, tormented way — to retain respect for human embryos while using an embryo or two for some crucial and momentary research purpose. Rather, we are contemplating an industry of embryo research — the production of countless numbers of embryos for use in research. Perhaps we shudder a bit at that prospect, but, if we do it for a while, it is likely that we will dull our sensibilities and learn not to shudder. Better to adopt, as Hans Jonas did in a classic essay on the ethics of experimentation, "the inflexible principle that utter helplessness demands utter protection."[1]

"Spare" Embryos

We might, though, shift the argument just a little. Suppose we say that we will show "respect" for human embryos in the following way: We

1. Hans Jonas, "Philosophical Reflections on Experimenting with Human Subjects," in *Philosophical Essays* (Englewood Cliffs, NJ: Prentice-Hall, 1974), p. 126.

will not create embryos simply in order to use and then destroy them in research. Rather, we will use only "spare" embryos — those originally produced for in vitro fertilization procedures but which were not needed or used (and remain frozen and all too "available"). These embryos, we might say, are destined eventually to be discarded and therefore have no future life prospects. Since they are destined to die, the only question is how. If we bring about their death by using them in research, could we not say that nothing is thereby lost (and it may be that something of use is gained)? Why not do this? Why not bring some good out of their deaths?

Such an approach has an initial attractiveness and may seem less crass than simply endorsing embryo research without limits. Some may support this position solely for strategic reasons — seeing it as the first but not, they hope, the last step in embryo research. But for others it may demonstrate a praiseworthy inclination to shudder just a bit in the face of a routinized use of embryonic human life for our own goals and purposes. Nevertheless, despite its initial attractiveness to some, there are serious reasons to turn aside from this line of argument. "Nothing will be lost, and something may be gained." It may be useful to ponder a similar kind of reasoning in two instances having nothing at all to do with research on embryos.

Perhaps the most well known example of research gone horribly awry in this country is the Tuskegee syphilis experiment (which lasted from 1932 to 1972). For approximately forty years, officials of the U.S. Public Health Service used impoverished, uneducated black men in Macon County, Alabama, as research subjects in a project designed to study the effects of untreated syphilis. Even after penicillin was known to be effective in treatment of patients with syphilis, these men were left untreated in order to follow the course of their disease. We miss some of the complexity of the case, however, if we forget that the poverty, illiteracy, and race of these men meant that, *even if the research had not been undertaken,* they almost surely would not have gotten treatment.

About those life circumstances — poverty, illiteracy, race — the Public Health officials could do little. Carrying out their research would not further diminish the life prospects of these men, nor would it impose upon them any additional risk of harm. (In fact, they no doubt received more medical attention than they would otherwise have gotten.) Why not, therefore, gain from their plight at least some useful knowledge that might benefit future sufferers? In his well-known book on this subject, James H. Jones describes (but does not endorse) precisely such reasoning: "The fate of syphilitic blacks in Macon County was sealed (at least for the immediate future) regardless of whether an experiment went forward. Increasing the store of knowledge seemed the only way to profit from the suffering there."[2] Nothing is lost, and something of potential medical significance is gained. Why not proceed?

Or consider the most notorious example of research ever conducted. When prisoners arrived at a concentration camp such as Auschwitz, "selections" were made that determined the life prospects of those prisoners. Many were simply fated to die. Discussing the way in which doctors at Auschwitz, "hungry for surgical experience," found these subjects ready to hand, Robert Jay Lifton writes: "In the absence of ethical restraint, one could arrange exactly the kind of surgical experience one sought, on exactly the appropriate kinds of 'cases' at exactly the time one wanted. If one felt Hippocratic twinges of conscience, one could usually reassure oneself that, since all of these people were condemned to death in any case, one was not really harming them."[3] Nothing is lost, and something of potential medical significance is gained. Nevertheless, surely we flinch from such reasoning here.

I do not contend that those who argue for research on "spare" embryos ought to be equated with the Tuskegee researchers or the

2. James H. Jones, *Bad Blood: The Tuskegee Syphilis Experiment* (New York: The Free Press, 1993), p. 94.

3. Robert Jay Lifton, *The Nazi Doctors* (New York: Basic Books, 1986), p. 295.

Nazi doctors. (Neither, of course, do I seek to relieve the conscience of anyone who may be bothered by the striking similarities in argument.) I simply wonder whether we should not be more troubled by the kind of argument used to support research on embryos produced for but not used for infertility treatments.

Indeed, the very fact that these embryos are leftovers from someone's attempt at IVF suggests not that they may now be used, but, rather, that they should not be used. After all, these embryos have already been used once in the service of someone else's project (possibly, we can grant, used justly in that project). They have been produced and used in an attempt to satisfy the desires of others. Is being used once not enough? Why, if they are no longer needed or wanted for reproductive purposes, should we suppose that they are still available for our use, still a handy resource for other purposes entirely unrelated to their well-being or their natural end? That they are destined to die anyway does not mean that we should feel free to use them but, rather, to quote Hans Jonas once again, that we should spare them "the gratuitousness of service to an unrelated cause."[4]

These embryos, we must not forget, are destined to die by our own will and choice. We cannot pretend that their dying is a natural fact unaffected by our choices. First we decide that they must die. Then we say that, since they're destined to die anyway, we might as well gain some good from that tragedy. It is true that — given certain choices that have been made — these embryos are destined to die, but our relation to their dying is not morally indifferent. It is one thing for us to acquiesce in their death; it is quite another for us to embrace that death as our aim, to seize upon it as an advantageous opportunity to use them yet again for our purposes. If we do that, something will surely be lost — something of great moral importance.

4. Jonas, "Philosophical Reflections on Experimenting with Human Subjects," p. 127.

Cloning Embryos for Research

Up to this point I have focused on the moral problem of using (and destroying) embryos in the course of research. An additional twist to our topic is provided, however, by the fact that many researchers would find especially useful the production of cloned embryos for experimental purposes. In this way they could produce precisely the disease models they want to study. And, as long as no one implanted — or permitted implantation of — these embryos, no cloned human being would be brought to live birth. There are, however, serious reasons to be concerned about such research goals. Here I note just two sorts of concern.

Suppose we wanted to produce cloned embryos for experimental purposes but, at the same time, wanted to ensure that no cloned human being was brought to live birth. How could we manage that? We would need to prohibit in our law the implantation and gestation of cloned human embryos. That is to say, we would have to *require* the destruction of a class of human embryos — to create a class of embryos that it would be a crime *not* to destroy. This would go well beyond our law in the matter of abortion — where government *permits* but certainly does not *require* the destruction of fetuses. We might want to think long and hard before putting our support behind a requirement of that kind.

A second concern takes the shape of a slippery slope argument. There is really no natural stopping point to this research, no good reason why it would stop with cloned *embryos*. For the moment, most advocates of such research argue that research should be limited to the first fourteen days of embryonic development and the embryo destroyed at that point. (As I noted in Chapter 3, this fourteen-day limit is going to seem increasingly arbitrary and unpersuasive in light of our knowledge of early embryological development.) What we could easily discover as research proceeds, however, is that embryonic and fetal organs may be more useful than embryonic stem cells. That is, there may be greater therapeutic benefit to be gained from harvesting differentiated tissue from early fetuses than from harvesting undifferentiated stem cells from the embryo. Why, after all, take stem cells and try to tease out of them differentiated lines of development when, if we wait a bit longer, nature herself will begin this work of differentiation for us? No one can say for sure that this will happen, but no one should be confident that the argument will not take exactly this direction. It may prove very hard to resist growing cloned human embryos to later stages in order to obtain more useful tissues for transplant. At that point there may not be much farther to slide down the slope.

Accepting Suffering

Medicine and medical research have a place of honor in the story of our attempts to relieve suffering. It is a goal whose nobility none of us should deny, a goal Christians are eager to honor and support. Yet, even our noble humanitarian projects do not always have moral trump, and we must evaluate not only the goals we seek but also the means to those goals. It is at least possible, therefore, that we might have to renounce some means to the worthy end of relief of suffering.

Discussing some sermons of St. Augustine, first preached in the year 397 but newly discovered in 1990, Peter Brown notes that Augustine was often required to preach at festivals of martyrs. At Augustine's time the cult of the martyrs — the "muscular athletes" and "triumphant stars" of the faith — continued to be of profound importance to ordinary Christians. Nevertheless, Brown suggests that in these sermons one can see Augustine quite deliberately underplaying the martyrs' feasts in order to emphasize instead God's everyday work in the heart of the ordinary believer. Those average Christians did not doubt the courage of the martyrs, but they questioned their own ability to accomplish anything even remotely as heroic in the fabric of their everyday lives.

In response, Augustine tells his hearers: "God has many martyrs in secret. . . . Some times you shiver with fever: you are fighting. You are in bed: it is you who are the athlete." And Brown comments:

> Exquisite pain accompanied much late-Roman medical treatment. Furthermore, everyone, Augustine included, believed that amulets provided by skilled magicians (many of whom were Christians) did indeed protect the sufferer — but at the cost of relying on supernatural powers other than Christ alone. They worked. To neglect them was like neglecting any other form of medicine. But the Christian must not use them. Thus, for Augustine to liken a Christian sickbed to a scene of martyrdom was not a strained comparison.[5]

It must have been a hard renunciation indeed; yet we see here a way of life for which relief of suffering — however greatly to be desired — is not the overriding imperative. We can learn from these Christians to break the hold on our own, understandable, tendency to believe that nothing can count for more than medical progress in the relief of suffering.

That god will fail us, and we must therefore break its hold on us before, like all idols, it breaks our integrity. We can do this only as we remind ourselves that, however greatly we value the betterment of life made possible by medical research, we have no overriding obligation to seek such betterment. Research brings betterment of our life; it does not save our society — or us. Noble goal that it is, medical progress is always optional, and, to cite Hans Jonas yet once again, there is "nothing sacred about it."[6]

5. Peter Brown, *Augustine of Hippo: A Biography,* new edition with an epilogue (Berkeley and Los Angeles: University of California Press, 2000), p. 454.

6. Jonas, "Philosophical Reflections on Experimenting with Human Subjects," p. 131.

CHAPTER TWELVE

Sickness and Health

Bioethics is not just about the difficult moral problems discussed in previous chapters, nor is it only about decisions we must make in one moment or another. More fundamentally, it invites us to think about the way we live toward death in a world marked by illness and suffering. And it should provide an occasion for us to consider how our way of life is shaped by the fact that we trust in a God who suffers for our redemption. Wherever we turn we are likely to confront sickness and suffering, and if we have been blessed with good health the disparity between our own condition and that of others may make the sheer fact of suffering all the more striking. We need to think about the meaning of illness in human life.

Responsibility for Illness

In the ninth chapter of the Gospel of John, Jesus heals a man who has been blind from birth. Jesus' disciples, seeking an explanation for the man's blindness, ask, "Who sinned, this man or his parents, that he was born blind?" (9:2). They assume, as Israelites of the time would ordinarily have assumed, that such blindness cannot have been entirely random. Someone must be responsible for it. After all, in ancient Israel the notions of ritual uncleanness or defilement had often connected sickness with sin. A person suffering certain afflictions was "unclean" and unworthy to participate in the religious life of the people. Jesus turns away from such a tight connection between sickness and sin. "It was not that this man sinned, or his parents," Jesus says, "but that the works of God might be made manifest in him" (v. 3).

We may be uncertain how to react to Jesus' response. On the one hand, it suggests that the distribution of illness may be harder to fathom than the disciples thought. No connection between sickness and sin can simply be assumed. When I am ill, I need not assume that God is punishing me, singling me out for retribution. If that relieves us of one worry, however, what are we to make of Jesus' suggestion that the man's blindness is by no means a random occurrence — that, in fact, the providence of God was at work in his blindness, fashioning the opportunity for Jesus to work a great "sign" by healing him? Perhaps all we can conclude is that the reasons for sickness may often be beyond our understanding. Jesus does not say that sin never results in sickness, nor does he say that it always does. He does not say that sickness never strikes at random without apparent reason; he says only that it did not in the case of the man whom he healed. He leaves us pretty much on our own to puzzle over this question — with, however, the clear warning that our ways are not God's, and his purposes may be beyond our comprehension. But he does one thing more. He gives to the sufferer the dignity of being united with him in his own suffering, and he gives to all of us the duty of attending to the sick, directing and freeing us thereby to show compassion to all who are ill.

Clearly, sickness is not always randomly distributed. Sometimes it is hard to deny that, because of our actions, we bear a certain responsibility for it. If for forty years I am a heavy smoker, there is no particular reason for surprise if I develop lung cancer. We can acknowledge this despite the obvious fact that someone else who had never smoked might also get lung cancer, or the equally obvious fact that another heavy smoker might not. Granting all those possibilities, the distribution of this illness is not simply random. Because our actions do to some degree affect our health, it would be a mistake to avoid entirely the language of responsibility here. Because no perfect correlation can be established, however, it would also be a mistake to suppose that our first duty is anything other than care for the sick, in whom one must learn to see the face of Christ.

Emphasizing too strongly our responsibility for health invites the false notion that we can be our own healers, that we create both our sickness and our health. Thus, for example, Thomas Droege moves from the helpful suggestion that we think of health not simply as the absence of disease but as a wholeness nurtured by the presence of God to the more questionable assertion that "healthy souls make for healthy bodies."[1] That assertion does not do justice to what Jesus says of the man born blind. More important, it can be spiritually destructive in the lives of those who are sick and who struggle against the temptation to see in their illness a sign of God's rejection.

We need simultaneously to retain a sense of responsibility for health and a spirit of compassion for those who are ill. The necessity and the difficulty of such a delicate simultaneity is written into the Bible itself. From the story of Job we learn how limited are our attempts to trace God's judgment within history. Job's "comforters" are not role models for us. Better that they should simply have been compassionate. But within the Wisdom literature of the Bible the Book of Proverbs exists as a kind of companion piece to Job, and it will not allow us to avoid altogether the language of responsibility. More confident about our ability to discern God's judgment in history than is the writer of Job, the sage in Proverbs says that those who turn away from Wisdom "shall eat the fruit of their way" (1:31). And by contrast, those who listen to Wisdom "will dwell secure and will be at ease, without dread of evil" (1:33). In health and in illness, the providence of God is at work in our

1. Thomas A. Droege, "Congregations as Communities of Health and Healing," *Interpretation* 49 (April 1995): 120.

world and our lives. Because it is God's providence at work, we can be confident that order and reason can be discerned even in the troubles and difficulties of life. The events of life will not be entirely opaque to our understanding. But faith never becomes sight, and belief in God's providence is belief in *God,* not in an impersonal process whose course and meaning we can master and control.

Hope in the Face of Illness

Sickness, suffering, and death are evils that plague human life, but they are not the greatest evil. That would be to lose God, to have reason to doubt his faithfulness to us. Christians affirm God's faithfulness, although they confess that it is demonstrated in a manner that might never have occurred to us had God not simply done it. God defeats and destroys the negative powers of sickness and death, but he does it by claiming even that realm as his own — by entering it and bearing it to its own logical end. The perfection and power of God is displayed in the acceptance of neediness, dependence, and even suffering. In the Epistle to Diognetus, an early Christian writing from the second or third century, the striking paradox of God's work in Jesus is forcefully expressed. How did God send his Son to overcome the world's destructive powers and to bring healing and life to us?

> Now, did he send him, as a human mind might assume, to rule by tyranny, fear, and terror? Far from it! He sent him out of kindness and gentleness, like a king sending his son who is himself a king. He sent him as God; he sent him as man to men. He willed to save man by persuasion, not by compulsion, for compulsion is not God's way of working.[2]

2. "The So-Called Letter to Diognetus," trans. Eugene R. Fairweather, in *Early Christian Fathers,* The Library of Christian Classics, vol. 1 (Philadelphia: Westminster

The true God will therefore always disappoint our desire for independence and self-sufficiency. Often what we want of that God is simply to make the powers of sickness and death disappear — and then leave us alone! But God wills to be with us in our dependence, teaching us, in William May's words, to overcome our "preoccupation with death and destructive power" and replace it with "attentiveness before a good and nurturant God."[3]

This need not and should not mean a rejection of the penultimate healing that scientific and clinical medicine offer us. The best physicians know, however, that their art at its highest must cooperate with powers beyond their own. We should give them our respect and our gratitude, but not our devotion — and they, of course, should seek no more. Instead, we place our ultimate hopes for Health and Wholeness in the God who himself has been broken by death — and who nevertheless lives. "Hope," G. K. Chesterton once wrote, "is the power of being cheerful in circumstances we know to be desperate."[4] That hope is the gift of God to us even and especially in our illness.

Press, 1953), p. 219. In *Suffering: A Test of Theological Method* (Philadelphia: Westminster Press, 1982), Arthur McGill used this passage to good effect in his discussion of God's paradoxical way of dealing with human suffering (cf. p. 82).

3. William F. May, *The Physician's Covenant* (Philadelphia: Westminster Press, 1983), p. 32.

4. G. K. Chesterton, *Heretics*, in *The Collected Works of G. K. Chesterton*, volume 1, ed. David Dooley (San Francisco: Ignatius Press, 1986), p. 125.

Index